The Auditory Cortex

The cover illustration shows neurones and glial cells of the auditory cortex of the marmoset monkey, stained with the Golgi stain.

The Auditory Cortex

Structural and functional bases of auditory perception

LINDSAY AITKIN

CHAPMAN AND HALL
LONDON • NEW YORK • TOKYO • MELBOURNE • MADRAS

UK	Chapman and Hall, 11 New Fetter Lane, London EC4P 4EE
USA	Chapman and Hall, 29 West 35th Street, New York NY10001
JAPAN	Chapman and Hall Japan, Thomson Publishing Japan, Hirakawacho Nemoto Building, 7F, 1-7-11 Hirakawa-cho, Chiyoda-ku, Tokyo 102
AUSTRALIA	Chapman and Hall Australia, Thomas Nelson Australia, 480 La Trobe Street, PO Box 4725, Melbourne 3000
INDIA	Chapman and Hall India, R. Sheshadri, 32 Second Main Road, CIT East, Madras 600 035

First edition 1990

Typeset in 10/12 Times by Mayhew Typesetting, Bristol
Printed in Great Britain by St Edmundsbury Press,
Bury St Edmunds, Suffolk

ISBN 0 412 32490 3

British Library Cataloguing in Publication Data

Aitkin, Lindsay
The auditory cortex.
1. Man. Auditory cortex
I. Title
612.8255

ISBN 0-412-32490-3

Library of Congress Cataloging-in-Publication Data

Aitkin, Lindsay.
The auditory cortex: structural and functional bases of auditory perception / Lindsay Aitkin. - 1st ed.
p. cm.
Includes bibliographical references.
ISBN 0-412-32490-3
1. Auditory cortex. 2. Auditory perception. 3. Cortical deafness.
I. Title
QP383.15.A35 1990 90-31869
612.8′255-dc20 CIP

This book is dedicated to Clinton N. Woolsey who, as a researcher and teacher, is one of the fathers of modern brain research, particularly that of the cerebral cortex.

Contents

Preface

There has been a proliferation of studies of the physiology of the sensory areas of the cerebral cortex in the last 30 years, a proliferation that stems in part from the Nobel-prize-winning work of Hubel and Wiesel on the visual cortex. Investigation of neuronal activity in the auditory cortex has been slower to develop, partly because the principles underlying its organization have been less easy to discern, at least at the single neurone level. In the 1970s the microelectrode revealed facets of the response properties and organization of the auditory cortex, and tracer substances were used to define the connections of cortical auditory areas. In the 1980s, methods for measuring functional activity and relating such activity to the anatomical substrate have been developed and applied to research on the auditory cortex in both animals and humans.

This account of the auditory cortex brings together research in the various disciplines of neuroscience. Its purpose is to give the reader an adequate background for understanding the mechanisms of normal auditory cortical function and cortical deafness. It is designed to inform readers with diverse backgrounds, but some postgraduate academic training, including physiologists, anatomists, psychologists and neurologists. To this end the terminology needed to express ideas about auditory cortical structure and function is related to a chapter in which methodology is stressed. Issues such as structure and functional and connectional anatomy lead into an account of the physiological properties of the auditory cortex, their ontogenetic development and their expression in the behaving animal and perceiving human.

I would like to thank my friends and colleagues Ramesh Rajan, Dexter Irvine, Jim Willott, and Rick Reale for their comments and advice about sections of this book. I also thank Dianne Clare, Frank Groeneveld and the resources of the illustrations laboratory of the Department of Physiology, Monash University. I am particularly grateful for the secretarial and editorial help of Valerie Park, who has put so much time into this work.

Lindsay Aitkin
Melbourne

'Nothing can be rushed
It must grow, it should grow of itself . . .
We must go on seeking it
We have found parts, but not the whole!'

PAUL KLEE
from 'On Modern Art'
Faber and Faber Ltd, 1948

1 An introduction to the auditory cortex

Humans suffering from middle cerebral arterial ischaemia, or from tumours of the temporal lobes, experience varying forms and degrees of 'cortical deafness'. These conditions occur following permanent damage to circumscribed regions within the temporal lobe. Such regions constitute the auditory cortex, which may operationally be defined as that part of the cerebral cortex concerned with hearing. For the purposes of this book we will restrict our coverage of hearing to the perception of sound and not include the interpretation of speech or the emission of sound. As a consequence, 'auditory cortex' will not include areas concerned with vocalization, such as Broca's or Wernicke's areas in humans. There is an enormous literature about human speech production and perception that cannot be covered here and very little of this material concerns the physiology of speech centres. Furthermore, it is difficult to make analogies and homologies between human speech and animal vocalizations. There is currently little evidence in non-human animals for specialized areas corresponding to Broca's and Wernicke's areas, although it is also true that very few relevant studies searching for these areas have been performed. A useful account of the anatomy of speech perception may be found in Seldon (1985).

The studies of Ferrier (1876), carried out over one hundred years ago, provided the first insights into the gross localization of the cerebral area concerned with hearing, and mark the beginning of the era of localization and structure that lasted beyond the inception of electrical recording methods. Ferrier employed electrical stimulation of the cerebral cortex in a number of species to show that excitation of temporal regions led to pricking movements of the opposite pinna and turning movements of the head and eyes. Bilateral lesions of this area in the monkey abolished responses to sound although the animal remained alert. Ferrier concluded (p. 175) that 'when the two sets of experiments are taken together, – viz., the positive reaction to electric stimulation, and the absence of reaction to usual forms of auditory stimuli when the superior temporo-sphenoidal convolutions were destroyed, – the evidence of the localization of the centre of hearing in this region amounts to positive demonstration'.

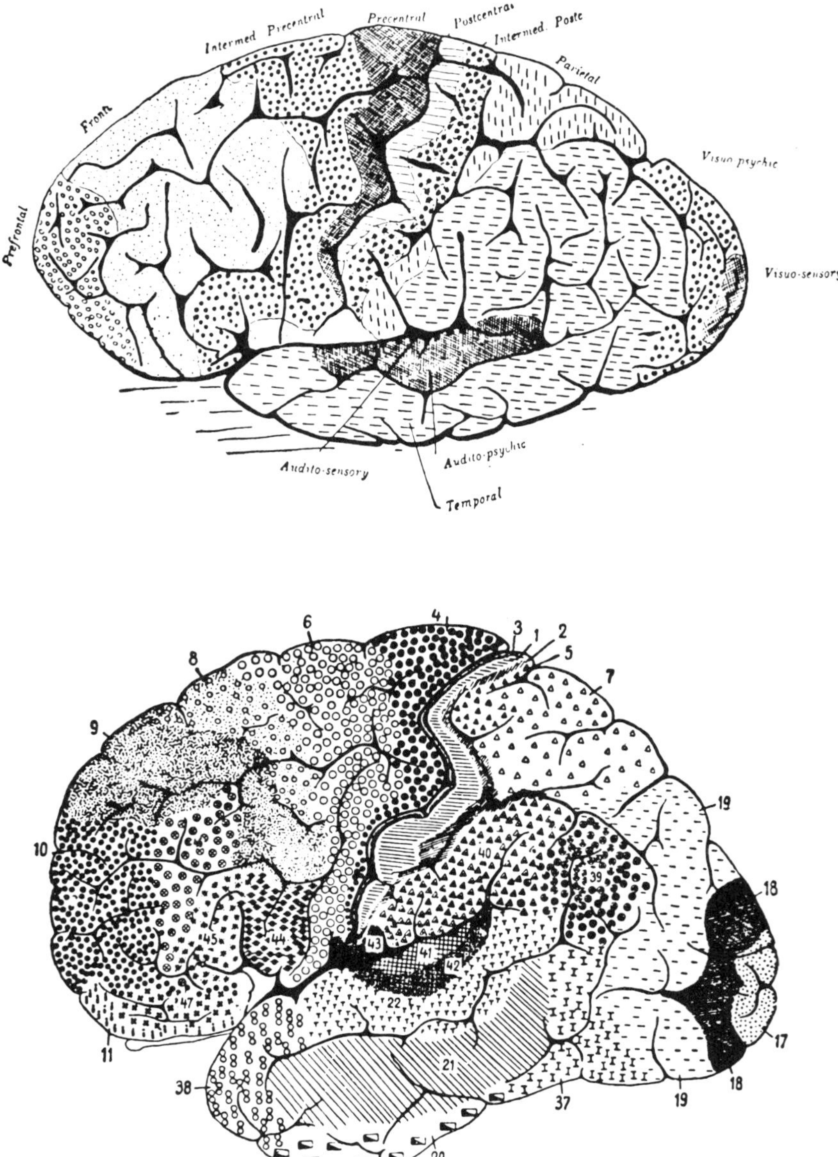

Figure 1.1 Two views of the human cerebral cortex seen from its lateral aspect. Above: Campbell (1905) differentiated between an audito-sensory and an audito-psychic area partly on the basis of earlier clinical and animal experimental data. Below: Brodmann (1909) avoided functional terms and devised the numbering system still in use today. Thus, numbers corresponding to Campbell's auditory areas are 41, 42 and 22.

The study of the cellular organization of the cerebral cortex (cytoarchitectonics), revealed with stains that are specific for neurones and glia, flowered in the first half of the century, particularly in the work of Campbell, Brodmann and von Economo. It was possible with this technique to describe the cytoarchitecture of the auditory cortex in those animals studied with electrical stimulation and lesion, and with this knowledge to predict the location of the auditory cortex in many animals, including people.

Cytoarchitectonic studies (e.g. Campbell, 1905; von Economo, 1929) distinguished two types of neurone – small, stellate granular cells and pyramidal cells, large and small. The arrangements of these cell types into the six layers characteristic of the cerebral cortex varies across the cortex. The sensory cortex, including the auditory cortex, appears to be characterized in most species by a wide superficial zone of mainly granule cells, forming layers II–IV, a cell sparse layer V containing some large pyramidal cells and a layer VI composed entirely of medium-sized granule cells (von Economo, 1929). Cortex of this type is referred to as granular cortex or koniocortex. A detailed analysis of the cytoarchitecture of the auditory cortex of the cat carried out by Rose in 1949 remains today perhaps the zenith of the Nissl method as it was applied to the auditory system (see Chapter 6).

Two of these cytoarchitectonic views of the human cerebral cortex are shown in Figure 1.1. That of Campbell (1905) (Figure 1.1, upper) is perhaps the earliest and includes a functional subdivision of auditory temporal areas, lying near the deep Sylvian fissure, and of occipital visual areas. The map of Brodmann (1909) (Figure 1.1, lower) is perhaps the most famous and his numeration of cortical zones is still in common use, not only in studies of humans, but also in animal experimentation. Thus areas 17, 18 and 19 correspond to primary and secondary visual areas, while areas 41 and 42 are subdivisions of the auditory cortex. Neither of these surface views can reveal the considerable extension of area 41 on to the superior surface of the inferior bank of the Sylvian fissure, observed in all primates. In humans this expanded superior temporal plane is called Heschl's gyrus. Adjacent and dorso-caudal to area 42, in areas 22 and 40, lies a region whose importance in speech perception was recognized over 100 years ago by Wernicke.

In the early part of the 20th century anatomists defined the major relay stations, or major regions of synaptic contact between axon and cell body or dendrites, in the auditory pathway – cochlear nucleus, inferior colliculus and medial geniculate body – using experimental techniques based on the degeneration of cell bodies and myelin sheaths when the axon of a neurone is damaged (see Chapters 2 and 3 for details). The final link in the chain, between auditory cortex and medial geniculate, was forged by Mettler (1935), Poljak (1932) and Walker (1938), and confirmed that areas 41 and 42 in primates and cats, as defined cytoarchitectonically, were indeed connected to the ascending auditory pathway.

The introduction of electrical recording methods in the 1930s made it possible to record, from the surface of the cortex, potentials evoked by sound or by electrical stimulation of the inner ear. These early studies were limited not only by the recording equipment (amplifiers, oscilloscopes and oscillographs) but also by the acoustic stimuli that could be produced. It was relatively easy to apply spark discharges from a thermionic valve circuit to a speaker (making a 'click') or to stimulating electrodes, but the generation of pure sine wave acoustic stimuli was still in its infancy and the frequency responses of speakers were limited to low-frequency ranges because of the mass of their moving parts.

Studies carried out in the 1930s were important in showing that the cortical field, across which click-evoked potentials could be recorded, correlated with the granular cortex defined in earlier cytoarchitectural studies. In particular, Bremer and Dow (1939) confirmed Kornmüller's observations (as cited in Bremer and Dow, 1939) that the architecture of the excitable region included a broad internal granular layer with cells arranged in clear radial columns. Shortly after the publication of Bremer and Dow's work, Woolsey and Walzl embarked on their study of the organization of the cortical potentials evoked by electrical stimulation of different parts of the osseous spiral lamina (Woolsey and Walzl, 1942). This work could be taken as the beginning of the era of electrical recording in the auditory cortex that is only today declining in importance, and will be reviewed in more detail in Chapter 4.

The area of auditory cortex identified in the studies of Bremer and Dow (1939) and Woolsey and Walzl (1942) was smaller than that considered putatively auditory by Ferrier (1876). For the purposes of this book a simpler, physiological, definition of the auditory cortex could be: those cortical regions activated by acoustic stimuli. Woolsey, in 1960, published a synthesis of all preceding work determining the locations of auditory cortex. His summary map shows a large proportion of the neocortex in cats activated by acoustic stimuli, ranging from areas exclusively auditory to association areas also influenced by stimuli of other modalities. Neocortical activation by acoustic stimuli may be even more widespread than Woolsey's map showed. For example, Morrell (1972) reported that cells in the visual cortex of cats may be activated by acoustic stimuli located in the appropriate region of space.

Penfield and co-workers (Penfield, 1958; Penfield and Perot, 1963; Penfield and Rasmussen, 1950) demonstrated that electrical stimulation of the anterior transverse temporal gyrus of Heschl (Figure 1.2) produced

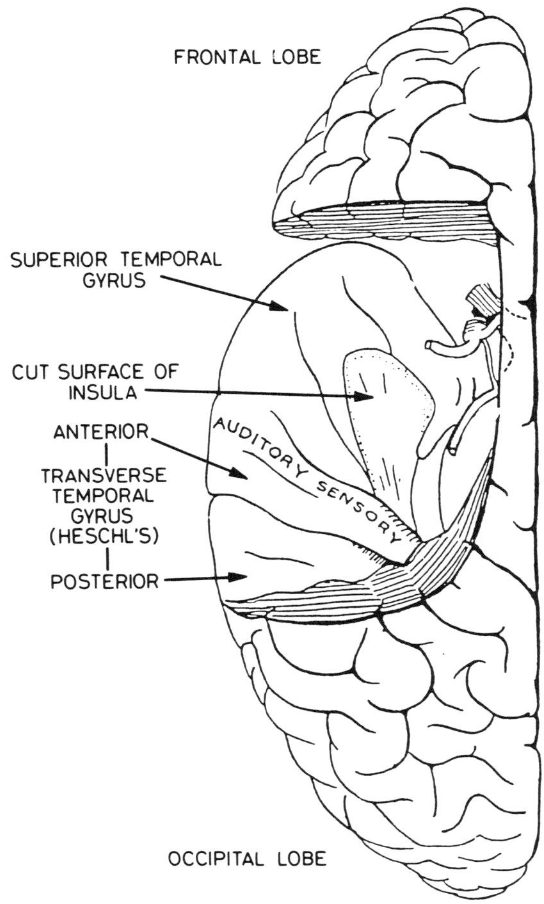

Figure 1.2 The left side of the human brain, viewed from above. Part of the dorsal surface of the cerebral cortex has been cut away to reveal the superior temporal and Heschl's gyri, where auditory and associated areas are located. Modified from Penfield and Perot (1963).

simple auditory sensations (buzzing, whistling, knocking) in conscious human subjects. Such perceptions also occurred when the temporal gyral sections of areas 41 and 42 were stimulated. Complex (experiential) auditory sensations occurred when rostral parts of the superior temporal plane and much of the dorsal, lateral surface of the temporal lobe were stimulated (Figure 1.2). There was thus some confirmation of Campbell's 'audito-sensory' and 'audito-psychic' cortex, and clear suggestions of 'primary' and 'secondary' auditory fields in the human brain.

It is obvious that a further restriction is needed in our definition of 'auditory cortex'. A series of classic studies by Woolsey, Walzl and Rose has provided a number of criteria that, when satisfied, distinguish an auditory cortical 'field' from association fields. First, each auditory field should have a complete and orderly representation of the cochlea. That is, stimulation using low-frequency tones, maximally effective at the apex of the cochlea, should cause activation of neurones in a part of an auditory field separated from the region activated by input from the base of the cochlea, evoked with high-frequency tones. The meanings of 'low' and 'high' are not absolute and will depend on the particular hearing range of the species studied. Neurones between these loci should show intermediate sound frequency preferences. Secondly, these fields should receive direct connections from, amongst other places, the medial geniculate body – the auditory part of the thalamus.

With this brief account in mind of the historical basis for what we will consider auditory cortex, let us now examine the way we shall study the auditory cortex by reference to the contents of subsequent chapters. The methods utilized to study the auditory cortex are examined in Chapter 2. Techniques for cell staining may be combined with the analysis of lesions and behaviour, with electrophysiology or with methods for tracing neural connections. Radioactive isotopes help in visualizing functional neural activity, especially, for the human brain, when used in association with positron emission tomography. The ability of radioactive emissions to reduce silver halides is important in autoradiography, a technique used in studying cell birthdays and for tracing neural connections. Certain dyes emit photons of different wavelength when the voltage in the medium in which they are dissolved changes magnitude. Such a property lends itself to studying changes in activity in a relatively non-invasive way. All of these methods and even more recent ones based on immunocytochemistry are now closely intertwined with computer technology, so that the computer often oversees, as well as analyses the results of, an experiment.

What is the nature of input to the thalamocortical auditory system? In Chapter 3 I outline the connections of the subcortical auditory system and examine the way information is modified and routed into parallel channels in its course from the cochlear nuclei to the auditory thalamus. At each level different synaptic processes modify the ascending information, leading to serial as well as to parallel integration.

In Chapters 4 and 5, the evidence for subdivision of the auditory cortex is examined from the point of view of cochleotopic organization (Chapter 4) and connections to the medial geniculate body and to other cortical regions (Chapter 5). The fine structure of the auditory cortex – the morphological varieties of cells and synapses, and the role of certain putative neurotransmitters – is related to the laminar and columnar organization of the auditory cortex in Chapter 6. Chapter 7 deals with physiological and functional aspects of the auditory cortex, such as the responses of neurones to stimuli of differing frequency and intensity. Comparisons between responses to the spectrally complex vocalizations of animals and to tonal stimuli will be made. In Chapter 8 I examine the representation of the two ears at the cortical level and evaluate the role of this region in sound localization and acousticomotor behaviour. The development of the auditory cortex, of its structure, physiology and function, is addressed in Chapter 9, in which the question of the plasticity of the auditory cortex of adults is also examined. In the last chapter I try to bring the various facets of auditory cortical organization together in an examination of the concept of cortical deafness as it relates to animals and humans. Here we attempt to relate our knowledge of neurophysiology and neuroanatomy to auditory perception by humans.

2 *Tools for studying the auditory cortex*

Our knowledge of the physiology of the brain stems from the application of a variety of techniques that are drawn from the basic sciences of physics (electricity, atomic physics, optics) and chemistry (dyes, enzymes, heavy metal compounds). We have already discussed briefly how the generation of an electric spark was used to activate regions of cortical tissue or segments of the cochlear nerve in the exposed cochlea, and to generate a 'click' sound from a loudspeaker. Some mention was also made of the development of amplifiers to enable the very small currents flowing extracellularly from excited neurones to be visualized on an oscilloscope.

I will examine a variety of techniques in this chapter that permit the study of the following questions. What are the discharge characteristics of auditory cortical neurones? These can be assessed by using microelectrodes to record the extra- and intracellular electrical potentials of cortical neurones. How can the morphologies of auditory cortical cells be determined, their arrangements into populations described, and their connections traced? What methods exist for relating these basic electrical and histological features to functional activity in behaving animals and perceiving humans?

MICROELECTRODE RECORDING TECHNIQUE

An important advance was made in the 1950s in the development of microelectrodes for recording from individual neurones (e.g. Ling and Gerard, 1949; Dowben and Rose, 1953; Green, 1958). These electrodes, being narrow tubes of electrolyte or metal surrounded everywhere but at the tip by glass or varnish, resist the flow of current and require special amplifiers for their successful use. Early devices utilized thermionic valves (vacuum tubes), but the science of electronics has grown considerably and there are now available many highly effective commercial amplifiers utilizing solid state circuitry.

Glass micropipettes may have tip diameters of less than 0.1 μm and are thus suited to intracellular recording from neurones (Figure 2.1a).

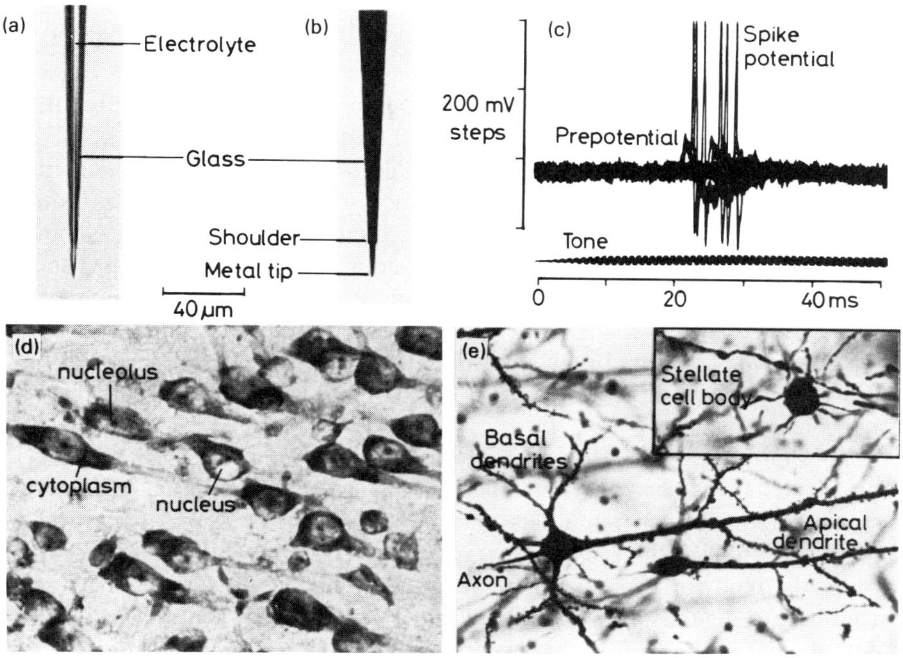

Figure 2.1 Aspects of some methods used in neuroscience research on the auditory cortex. Note that the size calibration is common for (a), (b), (d) and (e), so that some idea of the relative dimensions of cells and electrodes is apparent. (a,b) Glass and metal microelectrodes. (c) A typical oscilloscope trace showing a group of extracellular spike potentials evoked by a pure tone, the waveform of which is visible below. Some prepotentials, probably the extracellular manifestations of excitatory synaptic potentials, may also be seen. (d) Cells of the deeper layers of the temporal cortex of a cat stained for Nissl substance. (e) Golgi-stained pyramidal cell and stellate cell (inset). Axons, dendrites and dendritic spines are all visible.

They are usually filled with potassium chloride solution to match the most numerous intracellular cation species, and are used to measure resting, synaptic and action potentials of neurones. However, *in vivo* recordings from the cortex suffer from the problems of vascular and respiratory pulsations. These can be diminished, but rarely eliminated, with the use of a Davies chamber (see Erulkar *et al.*, 1956). The development of methods for recording from slices of cerebral cortex kept alive by superfusion of oxygenated physiological solutions may overcome problems of instability (e.g. Oertel, 1985).

An additional use for glass micropipettes is the injection of small quantities of dye or tracer into or around cortical neurones. The intracellular injection of fluorescent dyes or horseradish peroxidase (see pp. 12–13, 64) allows the morphology of the recorded neurone to be identified. Extracellular injection of horseradish peroxidase or other tracer substances permits the identification of the afferent and efferent connections of the micro-area from which recordings are made.

The tungsten microelectrode was introduced by Hubel in 1957 and in its various forms has proven the most successful extracellular electrode (Figure 2.1b). These electrodes are electro-etched in potassium nitrite solution to achieve a tip shape the dimensions of which are varied from one recording situation (e.g. recording from myelinated fibres) to another (e.g. insertion through the dura mater). Their strength is enhanced by insulation with glass and they can be used to make microlesions by passing a low positive direct current (2–10 μa) and causing a localized burn in the neural tissue.

Tungsten microelectrodes generally have lower impedances and lower noise levels than glass micropipettes, so that small extracellular currents associated with action potentials may be resolved from the noise in the recording system. However, only transient electrical changes (e.g. action potentials) tend to be recorded because steady potentials, such as resting membrane potentials, lead to a build-up of static charges (polarization) between metal and fluids that oppose current flow. The all-or-nothing extracellular action potentials (Figure 2.1c) recorded with these and other microelectrodes cannot be attributed unequivocally to the soma of a neurone but could be derived from the axon hillock, proximal dendrites or even from presynaptic or postsynaptic axons. The elements giving rise to these action potentials are thus referred to neutrally as 'single units'. The often considerable distances over which these units may be recorded from (e.g. Erulkar *et al.*, 1956) does argue, however for a somal origin. Recordings from myelinated axons are only likely to be obtained from a node of Ranvier, or if the electrode penetrates the sheath. Frequently the action potentials of a number of units can be recorded simultaneously; such recordings are considered to be derived from 'multi-unit clusters'.

CLASSICAL METHODS FOR NEURONAL AND FIBRE STAINING

Around 1880, Franz Nissl observed that certain aniline dyes adhered particularly strongly to subcellular components of neurones and glia. These 'Nissl bodies' are the 'ribosomes' of ultrastructural investigations and are composed of ribonucleic acid (RNA) to which the basic Nissl stains adhere. Two of the early dyes – thionine and cresyl violet – are still the dyes of choice for staining normal neurones (Figure 2.1d).

Ribosomal RNA is concentrated in clumps in the cytoplasm and nucleolus of a neurone, is dispersed in its nucleus and occurs in only small amounts in the dendrites or axon. Consequently only rarely can judgements about dendritic morphology be made from Nissl-stained material. In contrast, the method of heavy metal impregnation of neurones that bears the name of its discoverer, Golgi, can stain all of the processes of individual neurones but usually in only 1–2% of the population (Figure 2.1e). The exact histochemistry of this process is still unknown after 100 years, but it involves hardening the tissue in potassium dichromate, followed by immersion in silver nitrate or mercuric chloride, depending on the particular recipe. The soluble heavy metal salts are converted to oxides or metals that impregnate the neuronal processes, turning them black.

The Golgi technique is capricious, especially in relation to the staining of axons. These are more reliably stained with variants of the method of Weigert, which involves oxidizing myelin with a mordant such as potassium dichromate. In the presence of the mordant, vegetable dyes like haemotoxylin stain the myelin. However, the terminals and unmyelinated parts of axons cannot be visualized with this method and are usually stained with silver compounds which attach to the neurofilaments within the axon. The sites of silver deposition are revealed by a similar process to that used in photography – 'development' (reduction with ammonium hydroxide) and 'fixation' (with sodium thiosulphate). There are many variants of silver stains – all are generous in staining nearly everything in the nervous system and make it difficult to trace the paths of individual axons.

Another important use of silver compounds lies in their reaction to radioactive particles emitted by isotopes. Sections containing radioactive material can be covered by thin layers of photographic emulsions. The silver halides in these emulsions will be reduced to silver by radiation, and the silver can be revealed by photographic development and fixation. This process is called autoradiography. If an embryo is injected with tritiated thymidine, the nucleotide is taken up into deoxyribonucleic acid (DNA) by cells that are dividing at this time. Consequently the DNA itself will be labelled in the daughter cells, and only these cells, and can be detected for as long as the cells survive, by autoradiography. This forms the basis of determining neuronal 'cell birthdays'.

Finally, in ultrastructural investigations of the auditory cortex, heavy metal impregnation is necessary for electron microscopy. When tissue is illuminated by a beam of electrons the major species of atoms present in brain (C, N, O, H) do not deflect electrons sufficiently to form an image on a photographic plate. Instead the tissue is stained with heavy metals (osmium, uranium, lead) and these attach particularly to proteins, and so produce a non-selective image of most aspects of neuronal ultrastructure. The section thickness has to be about 0.06 μm to allow electrons to penetrate the tissue. The technique enables a high resolution so high magnification is possible. Cytoplasm is seen not as a homogeneous 'soup', but rather as a collection of distinctive organelles.

METHODS FOR TRACING THE CONNECTIONS OF NEURONES

Axons and cell bodies of the central nervous system cannot exist in isolation from each other. The neuronal soma supplies the necessary nutrients, building materials and transmitter substances (or precursors) that the axon needs, and the terminals can absorb materials from their immediate environment that are of use to the metabolic machinery of the neurone. Terminal and soma are linked by transport mechanisms present in the axon.

When the cell body is damaged, such as occurs following an experimentally produced lesion, the axon withers and dies. A dying axon has a greater affinity for silver compounds than a normal axon. This fact is made use of in the Nauta stains in which the tendency for normal neurofilaments to stain is suppressed with powerful oxidizing agents such as potassium permanganate. The result of these methods is that degenerating axons stand out sharply from relatively unstained tissue as black dots and dashes, and degenerating terminals may be seen as clouds of black dots. A similar rationale lies behind the earlier method of Marchi, in which degenerating myelin is visualized with haemotoxylin. This method, unfortunately, does not indicate where the degenerating axons terminate.

If all the terminals of an axon are damaged, the neurone will undergo retrograde degeneration. The neuronal soma increases its production of protein for a brief period and then loses ribosomal RNA activity and consequently its reaction to Nissl stains. Following this period of colour loss (chromatolysis) the soma dies and disappears. Rose and Woolsey (1949, 1958) employed this technique in their studies of the ascending connections of the auditory cortex.

These techniques all require the placement of a mechanical or thermal lesion to destroy cell bodies or axons. However, it is usually not possible to make lesions that are selective for a particular synapse; more often than not, other axons are damaged, including those *en passage*. This problem is overcome when use is made of the axonal transport processes of normal neurones. The horseradish peroxidase (HRP) technique has already been referred to in this chapter. This potent plant oxidative enzyme, when deposited in the vicinity of axonal terminals, is taken up by the terminals and transported back to the cell body projecting the axon. Its presence in the soma can be demonstrated by its ability to catalyse the oxidation of an intense dye (usually a benzidine derivative) from a colourless to a coloured form in the presence of hydrogen peroxide. Thus, only somas containing the HRP will become coloured when sections of the brain are incubated in the appropriate solution. The enzyme can also be used to trace anterograde projections, especially when it is conjugated with wheatgerm agglutinin.

Retrograde transport of fluorescent dyes, tiny plastic microspheres, and radioactively labelled transmitters, transmitter precursors or transmitter analogues can facilitate tracing and histochemically 'tagging' connections in the brain. In addition, axoplasmic transport of labelled amino acids is used to examine anterograde connections from neuronal somas. After microinjection in a brain region, the small labelled molecules are taken up by the soma and converted into labelled proteins. These in turn are transported to the axon terminals where their radioactivity may be detected by autoradiography.

The important soluble components of cytoplasm – those secreted by organelles in the cytoplasm, including transmitter substances and their enzymes – cannot be demonstrated directly with histological stains. Instead, techniques of immunocytochemistry are used in which antibodies to a particular enzyme are produced and the antiserum containing these antibodies applied to sections of the brain. The antibodies then bind to the enzyme against which they are raised and may be revealed with a secondary reaction that results, for example, in the complex becoming fluorescent or producing a colour reaction.

Antibodies may be produced readily to proteins, such as enzymes involved in the synthesis or degradation of transmitters, but the transmitter molecules themselves, being small, are not amenable to immunocytochemical techniques. However, it is possible to detect the presence of receptor molecules, to which the putative transmitter attaches, or of receptors important in the uptake of precursor molecules from the extracellular environment. Radiolabelled transmitter substances or small molecules specific for the receptors in question may be incubated with brain sections. After washing, the extent of receptor binding can be determined using autoradiography. These biochemical and immunological techniques, as they have been applied to the auditory system, are well reviewed by Wenthold (1985).

FUNCTIONAL ANATOMY

Functional anatomy may be defined as the morphological investigation of tissue that has been shown to have been activated by a stimulus. Two main principles form the basis for current demonstrations of activity in an anatomically distinct brain region. First, during increased metabolic activity in the brain the local cerebral arterioles dilate and there is an increased blood flow to the active area. Any substance normally uniformly distributed in the blood will show a high regional concentration in the active area. Secondly, metabolically active neural tissue has a high consumption of glucose. Metabolically inert homologues of glucose will be taken up by, and accumulate in, the respiring tissue.

Both these principles use the detection of radiation as their end-points. For the measurement of regional cerebral blood flow, increasing use is being made of positron emission tomography (PET). To quote Raichle (1983, p. 250) 'PET utilizes the unique properties of the radiation generated when positrons, produced by radioactive decay, are annihilated by interaction with electrons'. A positron-emitting atom such as ^{15}O or ^{11}C can be incorporated into carbon monoxide, and inhaled in small quantities by a subject. Carbon monoxide combines irreversibly with haemoglobin and the presence of the radioactive complex may be detected by the positron–electron annihilations that will occur everywhere, particularly in regions of high blood concentration. Each annihilation releases two high-energy photons that can penetrate the tissue for considerable distances in opposite directions. Detectors placed at opposite poles of the subject's head can detect the simultaneous occurrence of the high-energy photons and thus produce a 'picture' of regional blood flow in a certain cerebral plane. The sampling of radiation at various levels, or slices, of the brain constitutes 'tomography'.

Measurements of regional blood flow with PET or with inert radioactive substances such as xenon allow much information to be gained about the gross localization of function in the human cerebrum. For example, in Figure 2.2, Lauter *et al.* (1988) show how PET can reveal the simultaneous activation of different levels of the auditory pathway by, in this instance, synthetic syllables. The colour code indicates the magnitude of the regional cerebral blood flow; maximum is red to white. The top left slice is dorso-rostral to the superior temporal plane, in a cortical speech area, and shows a 16% left hemisphere advantage (see Chapter 10). The top right image is at the level of the superior temporal plane and presumably includes the primary and secondary auditory areas. The lower left and right 'slices' correspond to the medial geniculate bodies and inferior colliculi, respectively. These anatomical localizations are deduced from comparisons with a stereotaxic atlas of the human brain, and improvements of the method must incorporate more accurate localization methods.

More precise delineation of patterns of activity within an auditory region, such as the auditory cortex, is currently obtained mainly with histological tomography. Use is made of the observation that 2-deoxyglucose (2DG) is taken up by neurones as would be normal glucose, but it is not metabolized and accumulates in the tissue (Sokoloff *et al.*, 1977). If 2DG is made radioactive, its presence can be detected by the autoradiography of tissue sections, which are necessary because the radiation emitted by ^{3}H–2DG or ^{14}C–2DG can only penetrate for short distances. The histological nature of this process restricts the technique to animal brain tissue.

A new technique that may be amenable for use in humans as well as animals involves the demonstration of functional cerebral activity *in vivo* with voltage-sensitive dyes (e.g. Blasdel and Salama, 1986). The dye merocyanine fluoresces when small currents with time courses comparable to those of action potentials pass through the environment in which the dye is dissolved. The exposed cerebral cortex may be stained with the dye and optical signals emanating from patches of cortex as small as 100 μm in width may be sampled using optics and photodiodes or television tubes. In this way fine details of the localization of function across the cortical surface may be obtained with a relatively non-invasive technique. However, the acute and chronic toxicity of these dyes to the cerebral cortex remain to be determined.

It is likely that patterns of neural activity in the human auditory cortex will become resolvable using the magnetic properties of neural tissue. The movements of an electrical charge produce a magnetic field in the surrounding space. Many of these fields generated in the brain are so weak that they need a superconducting device to detect them. Such instruments are called superconducting quantum interference devices (SQUID) (Kaufman and Williamson, 1980). These have been complex, expensive and difficult to install, although less so in all respects than the cyclotrons needed to make the short-half-life isotopes for PET. However, in recent times considerable breakthroughs have been made in the field of superconductors, so it is possible that in the future the techniques of magnetoencephalography will become more accessible to laboratories of neuroscience.

One aspect of functional anatomical research has a long history – the deficits that occur in animal behaviour or human perception contingent upon deliberate or accidental lesions to the auditory cortical areas. Damage to the human cortical areas due to tumours or cerebrovascular insults will be discussed in detail later in this book – these deficits hardly qualify as 'tools' of neuroscience, more as tests of hypotheses. A traditional method for studying the functional roles of auditory structures of animals has been to observe the effects of lesions upon conditioned discriminations of, for example, sound frequency, intensity or position. From the pioneering work of Neff and co-workers (e.g. Neff, 1968) to the recent observations by Jenkins (Jenkins and Masterton, 1982; Jenkins and Merzenich, 1984), much is now known about the functional organization of the auditory system of cats and somewhat less about other species. The introduction of techniques for cortical mapping prior to behavioural manipulations or lesions has already revealed much about the plasticity of cortical areas in the adult (e.g. Kaas *et al.*, 1983).

Thus it is clear that the auditory cortex has been and is being studied in many species in many different ways. No longer can a student of cerebral function claim to be a neurophysiologist or a neuroanatomist, for example. A proper understanding of the auditory part of the cerebral cortex requires the integration of many facets of research in many disciplines.

3 *Basic features of the organization of the subcortical auditory pathway*

In this chapter we will briefly review the anatomy and physiology of the cochlea and provide an interpretive description of the subcortical auditory pathway. It is now apparent that parallel processing of information is a key principle in the design of the hindbrain auditory pathway; the following account is deliberately biased in this direction. Readers familiar with the physiology of the peripheral and subcortical auditory pathway may find it appropriate to pass directly to Chapter 4.

THE PHYSIOLOGY OF THE COCHLEA

The cochlea is a fluid-filled tube winding through the hard temporal bone that is divided internally into two main compartments by the cochlear partition. Components of the partition include Reissner's membrane, the osseous spiral lamina, the basilar membrane and associated sensory and supporting cells; these separate scala vestibuli from scala tympani (Figure 3.1). Scala vestibuli and tympani meet at the apex of the cochlea (the helicotrema); scala vestibuli receives incoming sound waves from the stapes at the oval window and scala tympani allows pressure changes in the fluid to be communicated with air through the round window. The pressure-release mechanism of the round window is important in allowing movement of the cochlear partition; otherwise the highly incompressible cochlear fluids and rigid bony capsule would not permit any movement of the stapes or support a travelling wave. Thus, inward movements of the stapes are coupled with outward movements of the round window membrane.

The non-uniform mechanical properties of the basilar membrane, together with an active mechanical process, are the basis of the mechanism of pitch discrimination. If the mechanical properties of the cochlear partition were constant between the base (where the stapes

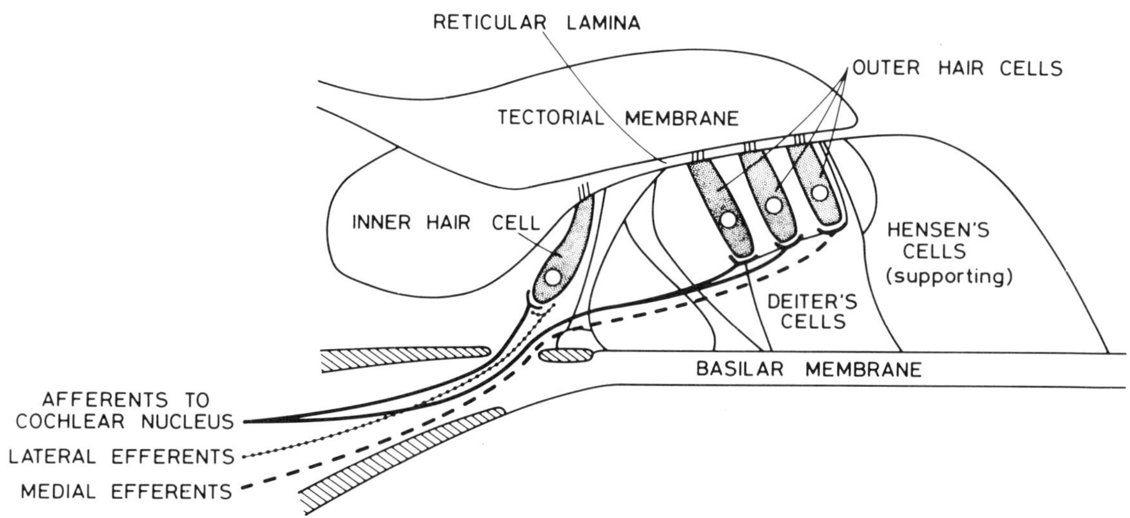

Figure 3.1 Schematic outline drawing of the organ of Corti to show in a simplified way the patterns of afferent and efferent innervation of the outer and inner hair cells.

inserts) and apex, then deflections of the partition would be of similar magnitude irrespective of location along the cochlea or of frequency of the sound. However, it was demonstrated many years ago by von Békésy (1960) that the thick, narrow basal part of the cochlear partition vibrated most strongly to high sound frequencies, whereas the wider, more flexible apex responded to low frequencies. Von Békésy enunciated the concept of a wave that travelled along the cochlea causing the greatest deflection of the cochlear partition at a region where the resonant properties of the basilar membrane matched the frequency of the sound wave. Von Békésy thus showed that there was a *tonotopic* organization along the basilar membrane.

The cochlear partition is a complex structure, being composed of the endolymph of scala media, the Reissner's membrane, the basilar membrane and other connective tissues (tectorial membrane, reticular lamina, rods of Corti) and a variety of epithelia, including the transducing hair cells (Figure 3.1). One major change in our thinking about mammalian cochlear mechanics, originating from ideas put forward by Gold (1948), is that vibration of the cochlear partition is an active process and that the sharpness of tuning of the cochlear partition – the length of the region vibrated by a specific frequency – is severely reduced post mortem or after noise damage, when outer hair cells are lost (for further information, see Pickles, 1982). It is currently considered that the sharp tuning of cochlear neural elements is due entirely to that of the cochlear partition.

Elements in the transduction process are identified in Figure 3.1. A single row of inner hair cells receives approximately 95% of the cochlear afferents, with each inner hair cell receiving up to 20 afferent fibres. Each afferent fibre innervates only one inner hair cell. The remaining 5% of afferents are distributed to three to four rows of outer hair cells – one cochlear nerve afferent can supply ten outer hair cells (Spoendlin, 1972). These outer hair cell afferents appear to be unmyelinated and have distinctive cell bodies in the spiral ganglion. Because of their small diameter, they may rarely have been recorded from in electrophysiological experiments (Kiang *et al.*, 1982).

Efferent fibres also supply the hair cell rows, and their patterns of connectivity have been studied in cats and guinea pigs. Efferents to inner hair cells terminate mainly upon the afferent dendrites and originate from neurones found in the lateral part of the superior olivary complex (Warr and Guinan, 1979). Efferent axons to the outer hair cells terminate directly upon those cells at cholinergic synapses and originate from neurones in the medial part of the superior olivary complex.

The dense afferent innervation of inner hair cells argues strongly in favour of their having a primary role in transduction. It is likely that deflection of the hairs or stereocilia of inner hair cells depends on the velocity of movement of the surrounding fluid at low frequencies, and becomes entrained to basilar membrane displacement at higher frequencies. Such movements are in some way translated into changes in the permeability of the membranes of the inner hair cells to ion species which differ in concentration on either side of the membranes. Resultant ion fluxes, it is believed, depolarize the hair cell (receptor potential) and cause the release of a neurotransmitter (perhaps an acidic amino acid) from the hair cell base on to the cochlear nerve terminals that synapse with the hair cell. The neurotransmitter depolarizes the terminal and the resultant generator potential leads ultimately to action potentials in the cochlear nerve fibre.

There are a number of independent lines of evidence to suggest that outer hair cells may have the ability to change their length and in so doing to alter the micromechanics of the cochlear partition (Brownell *et al.*, 1985). It is not clear how this property relates to the attachment of the cilia of outer hair cells to the overlying tectorial membrane, nor is there any single, generally accepted theory of their role in hearing. Some theories include the concept of feedback from the central auditory system, while others postulate direct electrical interactions between outer and inner hair cells, obviating the necessity for feedback and allowing extremely rapid adjustments of excitability to occur. Clinical evidence, and behavioural evidence in animals, indicates that substantial hearing losses occur when the outer hair cells are damaged. Thus, whatever the mechanisms the outer hair cells must have a vital role in augmenting the transduction capabilities of the inner hair cells.

PHYSIOLOGY OF COCHLEAR NERVE AFFERENTS

Afferents from the inner hair cells form a homogeneous group in terms of their physiological properties (Evans, 1981; Kiang, 1984; Rose *et al.*, 1967). All have some degree of spontaneous activity and the rate of spontaneous firing appears to be correlated with the range of suprathreshold sound intensity over which increases in firing rate occur. The origin of the cochlear nerve afferent along the cochlear partition is indicated by the 'characteristic frequency' (CF) of that fibre – the frequency requiring lowest sound pressure for excitation of the fibre. Each fibre, when activated by tone bursts, discharges at first very rapidly and then at a lower rate which is sustained for the duration of the stimulus. Both the peak and steady rates depend on the sound intensity and a maximum firing rate occurs for each fibre at a certain intensity. At sound pressures beyond this intensity the firing rate is said to be

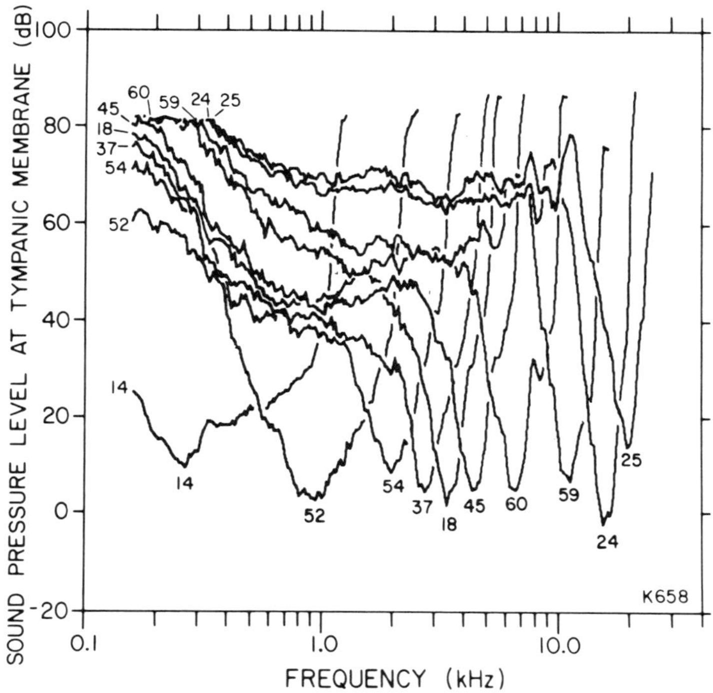

Figure 3.2 Tuning curves for ten fibres, each numbered, from the auditory nerve of a cat (K658). From Kiang (1984).

saturated. Finally, the temporal spacing of action potentials in a given nerve fibre depends partly on the intervals between peaks of sound pressure variations which, for a pure tone, are equal to the period of the tone.

Inner hair cell afferents demonstrate the sharp frequency selectivity displayed by the hair cells with which they synapse. Examples of threshold tuning curves – plots of sound pressure needed to evoke a threshold response as a function of tone frequency (Figure 3.2) – show that thresholds rise steeply on each side of the CF but at lower frequencies the rise becomes more gradual leading to 'tails' in the tuning curves of many cochlear nerve fibres.

The CFs of cochlear nerve fibres are related systematically to the position of the nerve fibre in the whole cochlear nerve in the modiolus. Fibres from more basal (high-frequency) parts of the cochlea, on joining the central nerve bundle, are located around those from the apex, giving the whole nerve its cochleotopic (and thus tonotopic) organization.

Simultaneous presentation of more than one tone, or presentation of complex sounds, may lead to the suppression of a response of a cochlear nerve fibre to a normally excitatory single tone. This phenomenon ('two-tone suppression') is probably due to complex mechanical interactions along the cochlear partition rather than to neural inhibition. There is no evidence to suggest that afferents from one group of hair cells directly inhibit the firing of other afferents, although a feedback loop involving the lateral system of efferents may inhibit the firing of inner hair cell afferents. The normal stimuli to provoke this feedback inhibition are presently unknown.

It is generally accepted that movements of hair cell stereocilia in one direction, but not the other, cause hair cell depolarization and excitatory transmitter release on to cochlear nerve fibres. Since these stereocilia movements will have a fixed phase relation to the movement of the basilar membrane, so too will the action potentials in cochlear nerve fibres. Thus, information about sound frequency is also contained in the temporal spacing between action potentials in a given cochlear nerve fibre. This phenomenon, called phase locking, can be observed up to frequencies of 4–5 kHz in mammals and even higher in the barn owl (Sullivan and Konishi, 1984). In addition to a possible role in spectral analysis, phase locking may be important in the central analysis of interaural time differences. Low-frequency sound waves from a single source in space will evoke phase-locked trains of impulses in homonymous cochlear afferents. The time difference between corresponding impulses will relate to the difference in time between wavefronts arriving at each ear – the interaural time difference. Thus, at the central site of convergence of these phase-locked trains, information about the location of the sound source in space may be gained by cross-correlation.

Increases in sound pressure above threshold lead to increases in firing rate over a 'dynamic' range of intensities that, for the majority of fibres, is restricted to 20–50 dB. For a minority (particularly those with low rates of spontaneous activity) the dynamic ranges may exceed 60 dB. It is likely that the coding of increases in intensity involves a number of factors, including the recruitment of fibres with similar dynamic ranges but higher thresholds and the decrease in spike latency accompanying such increases in intensity.

The representation of complex sounds, including vocalizations, is also likely to involve many factors. These include the changing firing rates of individual fibres tuned to the pure tone components of the complex sound, the varying recruitment of different populations of fibres as the sound frequency and intensity vary, and the fluctuating time relationships between successive action potentials of a given fibre and between corresponding action potentials in different fibres. Feature extraction, involving the emergence of neurones activated by a specific combination of acoustic features, is a more central function derived initially from the synaptic specializations that are present in the nuclei of the auditory brain stem.

THE ORGANIZATION OF THE BRAIN STEM AUDITORY PATHWAYS

A simple scheme of the wiring of the brain stem auditory system for a generalized mammal is shown in Figure 3.3. The connections at the medullary level, between cochlear and superior olivary nuclei, are shown in the lower part of the figure. Connections ascending to the midbrain (inferior colliculus and nuclei of the lateral lemniscus) are shown in detail in the centre, while the auditory thalamocortical system is sketched in above for completeness. The details of the latter part of the auditory system are, of course, a major part of this book and will be provided in later chapters. Each cochlear nerve fibre bifurcates upon entering the cochlear nuclear complex. A posterior ascending branch supplies cells in the posteroventral (PVCN) and dorsal (DCN) cochlear nuclei and an anteriorly directed branch innervates the anteroventral cochlear nucleus (AVCN). Each of these three major divisions has a complete tonotopic organization and each contains a number of morphologically distinct cell types. The physiological properties and efferent projections of some of these cell types have been studied in detail, particularly in cats (see Irvine, 1986).

Each 'bushy' cell of the AVCN has a few short stubby dendrites and receives input from only a small number of cochlear nerve fibres, principally in the form of large axosomatic terminals: end bulbs of Held. Because of the powerful excitatory synaptic input they receive, bushy cells faithfully transmit information about the timing of cochlear nerve action potentials. They project strongly to the superior olivary complex (Figure 3.3, Pathway A).

'Stellate' cells are found in both the AVCN and PVCN and receive their cochlear nerve input axodendritically from boutons terminaux of many cochlear nerve fibres. Their physiological properties are compatible with spatial synaptic integration, perhaps at the expense of temporal information. They project predominantly to the auditory midbrain and nuclei of the lateral lemniscus of the opposite side (Figure 3.3, Pathway B). Another type of neurone in the PVCN, the octopus cell, has long dendrites that receive an even more convergent cochlear nerve fibre input, with the resultant dilution in the representation of any one fibre.

There is a general correlation between the firing patterns of bushy,

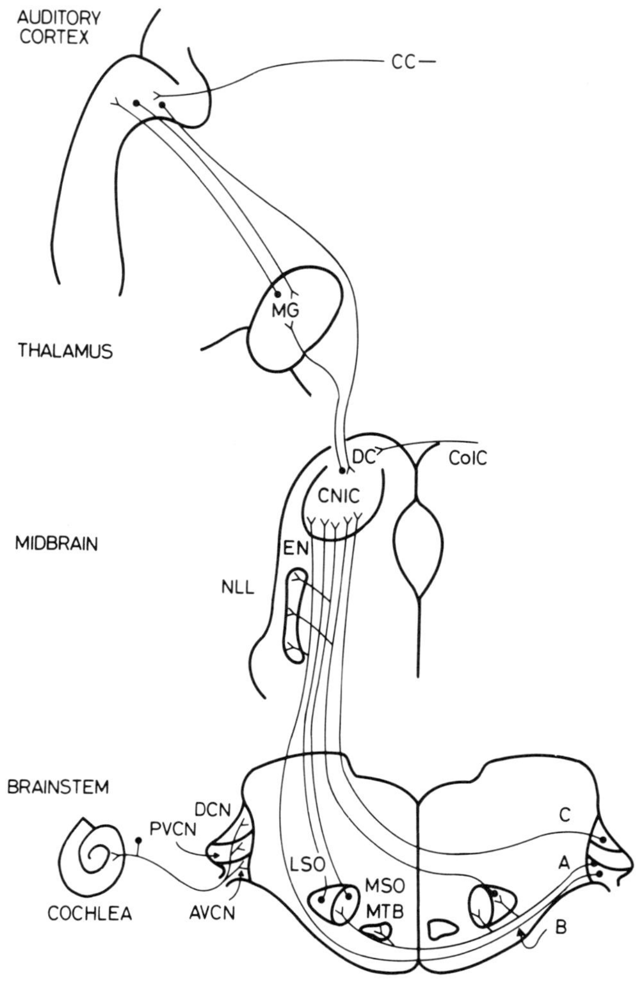

Figure 3.3 The parallel streams of information flowing through the brain stem of a mammal. Pathway A relays in the anteroventral cochlear nucleus (AVCN), medial (MSO) and lateral superior olive (LSO), and medial nucleus of the trapezoid body (MTB) before terminating in the central nucleus of the inferior colliculus (CNIC). Collaterals pass to the nuclei of the lateral lemniscus (NLL) *en route*. Pathway B also originates from the AVCN (some cells in the posteroventral cochlear nucleus, PVCN, may also be involved) and crosses directly to the CNIC, as does pathway C, emanating from the dorsal cochlear nucleus (DCN). Some of these crossed projections may terminate in the external nucleus of the inferior colliculus (EN). The core pathway continues to the medial geniculate body (MG) and auditory cortex. The latter also receives input from the corpus callosum (CC), and projects back to the MG and to the inferior colliculus, particularly the dorsal cortex of the inferior colliculus (DC). Axons of the commissure of the inferior colliculus (ColC) also terminate in DC.

stellate and octopus cells and the density of terminations from individual cochlear nerve fibres: bushy cells have firing patterns very similar to those of cochlear nerve fibres ('primary-like'), stellate cells fire repetitively but with small temporal gaps between bursts of action potentials ('chopper'), and octopus cells fire only at the onset of the tone burst.

Input to the dorsal cochlear nucleus is derived as much from descending sources as from the cochlear nerve. The layered structure of this nucleus in many species has been compared to the cerebellar cortex (Osen and Mugnaini, 1981). The principal projecting neurones are the pyramidal cells, whose axons terminate mainly in the contralateral auditory midbrain (Figure 3.3, Pathway C). Giant cells, found deeper in the DCN, also project contralaterally. Complex interactions of an excitatory–inhibitory nature occur between the small granule and large pyramidal cells in this nucleus. The responses of neurones of the DCN to tone bursts frequently show inhibitory pauses after a strong onset excitation, before the resumption of sustained excitation ('pauser' patterns). Other cells lack the onset response and have a 'build-up' pattern.

These correlations between morphology, physiology and efferent projections have suggested that the cochlear nerve fibre input is distributed to a number of parallel information channels in the cochlear nucleus. In this way different aspects of the discharges of cochlear nerve fibres are singled out by a process of synaptic 'filtering' in the cochlear nucleus. It is instructive to pursue the concept of parallel processing further into the relationships between the bushy and stellate cells of the AVCN, and the superior olivary complex.

PARALLEL PROCESSING IN THE AUDITORY BRAIN STEM

Auditory information flow can be followed over a number of routes that are largely separate (Aitkin, 1986). Several main streams connect the ventral cochlear nucleus with the inferior colliculus. Bushy cells of the AVCN project via the ipsilateral lateral superior olive (LSO) or the medial superior olive (MSO) bilaterally. Modified bushy cells of the AVCN project to the contralateral medial nucleus of the trapezoid body (MTB) which in turn projects to the LSO of that side. The LSO and the MSO are thus binaurally connected. The LSO projects to the central nucleus of the inferior colliculus (CNIC) of both sides whereas the MSO projects predominantly ipsilaterally to the CNIC. Subsidiary pathways from the superior olivary complex relay through the dorsal nucleus of the lateral lemniscus whose efferents project to the CNIC bilaterally and to the superior colliculus.

In addition to these binaural pathways, exclusively contralateral projections may be traced from pyramidal and giant cells of the DCN (Figure 3.3, Pathway C) and from stellate cells of the ventral cochlear nucleus (Figure 3.3, Pathway B). Both pathways terminate predominantly in the CNIC but a subsidiary pathway to the CNIC relays through the ventral nucleus of the lateral lemniscus. It is possible that connections to the nuclei of the lateral lemniscus are collaterals of pathways ascending directly to the CNIC. Although possibly subsidiary in this sense of receiving collaterals, the nuclei of the lateral lemniscus are nonetheless a major source of input to the central nucleus of the inferior colliculus. Note that only two pathways through the ipsilateral superior olive provide excitatory connections to the CNIC from the ipsilateral cochlear nucleus.

These parallel pathways terminate in different, although overlapping, parts of the inferior colliculus. It is possible to recognize the nuclear sources of input to a given neurone in the CNIC by comparing its properties with those of the neurones that lie along its afferent supply. In this way the extraction of information and the integration of this extracted information in the auditory midbrain may be analysed for these various pathways.

Pathway C, from the DCN, contains neurones whose properties are shaped markedly by excitatory and inhibitory processes, activated by acoustic input to one ear, that interact in the frequency and intensity domains. Their frequency response profiles show evidence for below-CF and above-CF inhibitory sidebands, and their responses as a function of intensity are usually sharply non-monotonic (firing rate increases, and then decreases, as stimulus intensity is increased). Pathway B, from the ventral cochlear nucleus, preserves some of the features present in the auditory nerve. Its neurones fire in a sustained fashion throughout a tonal stimulus, and both monotonic and non-monotonic intensity functions have been identified; all neurones are excited in a sustained fashion from the contralateral ear, and ipsilateral stimuli rarely have any influence.

Much is known of the processing occurring along binaural pathways at the level of the CNIC. Some neurones are excited by stimuli presented to one ear and inhibited by stimuli presented to the other in a manner dependent on the interaural intensity difference (Boudreau and Tsuchitani, 1968; Irvine, 1986). At least a proportion of these units are implicated in the coding of the location of high-frequency stimuli. For other neurones this feature, extracted by processing in the LSO, is combined with information carried along monaural pathways. The discharges of such neurones may encode both location and frequency–intensity characteristics of an acoustic stimulus. Neurones of the CNIC that receive input exclusively from the ipsilateral LSO are rarely found and it is likely that many of these neurones also receive input from pathways originating from the contralateral ear. Such units would exhibit binaural excitatory properties beyond the inferior colliculus.

The binaural integration of temporal information appears to be a principal function of pathways relaying in the MSO. Neurones in this structure and at higher levels are influenced by variations in the time difference between binaural stimuli and in some cases may show phase-locked discharges of the kind observed in the auditory nerve, as described previously in this chapter (e.g. Goldberg and Brown, 1969). It is thought currently that some of these neurones have properties appropriate for the coding of the location of low-frequency-containing stimuli. Other neurones of these pathways may be concerned with spectral analysis of low-frequency stimuli by binaural integration.

In addition to being a relay for these pathways en route to the cerebral cortex, the central nucleus of the inferior colliculus is the source of two other, qualitatively different, streams of information (Aitkin, 1986). Auditory information is channelled to the external nucleus of the inferior colliculus (Figure 3.3, EN) and from thence to a number of premotor areas, including the superior colliculus, reticular formation and cerebellum. These pathways are important in acousticomotor reflexes and rapid orienting responses. A separate circuit involves the dorsal cortex of the inferior colliculus (Figure 3.3, DC) and the commissure of the inferior colliculus (Figure 3.3, CoIC), feeds auditory information to more rostral non-specific areas, and may be concerned with attention and arousal mechanisms.

THE RESPONSE PROPERTIES OF COLLICULAR NEURONES PROJECTING AXONS INTO THE THALAMOCORTICAL AUDITORY SYSTEM

Let us now consider the patterns of activity, evoked by sound, in neurones whose fibres terminate in the medial geniculate body – the thalamic auditory centre. We will focus on those characteristics that are also well studied at the auditory cortical level. This will enable us to examine how such response features are re-represented at the cortical level and thus provide insights into the integrative mechanisms of the auditory thalamocortical system. The following summary is compiled from a number of sources, including Aitkin (1986), Calford *et al.* (1986), Merzenich and Reid (1974), Oliver and Morest (1984), Rose *et al.* (1963) and Semple and Kitzes (1985).

Pure tones excite most neurones in the inferior colliculus. Those in its central nucleus (CNIC) respond with activity at the onset of a tone burst stimulus. Additional spike activity is sustained for the duration of the stimulus, usually after a short pause. Neurones in the external nucleus (EN) or dorsal cortex (DC) are more likely to respond only at the onset of the stimulus and to prefer more complex or frequency-modulated stimuli.

Tuning curve shapes vary in sharpness over a continuum in the CNIC, while those in the EN and DC tend to be very broad. Presumably the latter neurones have lost frequency selectivity in a trade-off for other stimulus features – in the case of some neurones of the EN, somatosensory input. There is little doubt that frequency selectivity is an important feature of neurones of the CNIC, but only of a subset, and only in this subdivision of the inferior colliculus.

There is a very precise arrangement of neurones of the CNIC in relation to their CFs, in all species studied for this feature. Neuronal CF is relatively constant in one (isofrequency) dimension – such isofrequency 'planes' may, however, be quite complex in curvature – and changes rapidly in an orthogonal direction. There is some suggestion that this tonotopic organization parallels, and may correlate with, an anatomical arrangement of ascending fibre terminals and the dendrites of the principal or disc-shaped cells of the CNIC. This arrangement gives the CNIC a layered or laminated appearance in Golgi material.

Changes in firing rate as a function of intensity are monotonic or non-monotonic in approximately equal proportions of neurones in the CNIC. However, one could argue for a continuum in the neural coding of intensity between functions that become non-monotonic at intensities only 10–20 dB above threshold (at higher intensities the neurone is completely 'silent') to those that are monotonic over an 80–100 dB intensity range.

The collicular outflow is dominated by excitatory input from the contralateral ear. In dichotic studies contralateral activity is generally more vigorous, of lower threshold and of shorter latency than that evoked by equivalent ipsilateral stimuli. However, the majority of neurones (60–80%) in the CNIC are influenced by binaural stimuli and for some neurones the inhibitory effects of ipsilateral stimuli are at least as powerful as the excitatory contralateral input, on an equal intensity basis. Approximately one third of the units in CNIC show excitatory–inhibitory (EI) effects, varying from evenly matched interactions to weak effects in which the ipsilateral stimulus has to be much stronger than the contralateral stimulus to show any influence. Such weak effects suggest the possibility that monaural and EI units are extremes of the same population; which characteristic prevails may be determined by the density of afferent connections to a given neurone from monaurally or binaurally innervated brain stem nuclei. A further one-third of units in the CNIC show evidence for binaural excitation. In some units this is only revealed by varying the interaural time or intensity difference. Binaural stimuli presented with zero time delay or at equal intensities may evoke no more activity than the contralateral stimulus alone.

Some of these binaural units are likely to be involved in the neural coding of sound location. In studies employing free-field stimuli, units in the inferior colliculus have been classified as omnidirectional (no effect of different sound source locations) or directional (Aitkin *et al.*, 1984a). The latter grouping encompasses units varying from weakly directional (almost omnidirectional) to those whose firing rates are sharply dependent upon sound source location and are relatively independent of sound intensity. Such 'azimuth-selective' neurones seem well suited to have a role in sound localization. Comparisons with dichotic studies show that about 50% of the interaural-time-delay-sensitive units and about 25% of the EI cells are likely to be azimuth selective, so that it is possible that many binaural neurones may have only a secondary (if any) role in sound localization.

Most of the above observations have been made using pure tone stimuli. Such stimuli have the advantage of being perceptually simple, and the vibration patterns they evoke on the cochlear partition are well understood. Clearly, however, pure tones are unusual in nature and it would be of interest to know how biologically relevant sounds are encoded by brain stem mechanisms. Such information is lacking for the midbrain; more is known at the cortical level (Chapter 7). There is some information about the responses of neurones of the CNIC to broadband noise stimuli, particularly in binaural stimulation and free-field experiments. In dichotic studies the response of a neurone to a broadband noise stimulus can be predicted from knowledge of that unit's responses to pure tone stimuli (Geisler *et al.*, 1969; Yin *et al.*, 1986). In the free field some units are sharply directional to noise stimuli yet are omnidirectional to tonal stimuli (Aitkin *et al.*, 1984a; Aitkin and Martin, 1987). This suggests that there are complex interactions between a broadband noise stimulus and the head and pinna, probably resulting in some spectral transformation of the noise that may be different for the two ears. As a consequence, the two cochleas may be 'seeing' two spectrally different noise stimuli in the free field but not in the dichotic situation.

This description of the discharge characteristics of neurones in the CNIC has highlighted the binaural connectivity that is an important feature of integration in this nucleus. It is now of interest to follow the destinations of axons of the inferior colliculus into the auditory thalamus, where further synaptic integration is modulated by local and corticofugal input prior to relay to the auditory cortex.

THE PRESERVATION OF PARALLEL PROCESSING CHANNELS IN THE AUDITORY THALAMUS

Axons leaving the inferior colliculus ascend to the medial geniculate body (MG) along the brachium of the inferior colliculus. Most axons project to the ipsilateral MG but a small proportion pass across the commissure of the inferior colliculus to the contralateral MG. An adjacent part of the thalamus, the lateral part of the posterior group of thalamic nuclei, Po, also receives collicular input. Together with the MG, these nuclei constitute the auditory thalamus.

The cytoarchitecture of the MG has been described in a number of species on the basis of Nissl-stained material and cellular morphology revealed with Golgi stains. The most detailed analysis has been made in cats, principally by Morest (1964, 1965) and by Winer (1985a). There is a tendency for descriptions of the MG of other species to be linked to those of cats, but there is no necessity that the feline MG should be prototypic, and some of the smaller subdivisions recognized in cats are difficult to discern in, for example, primates and marsupials.

If anatomical subdivisions and physiological properties are related to connectional data, certain features of the organization of the MG appear similar across those species for which comprehensive data are available. For convenience, the nomenclature of Morest will be used (dorsal, ventral and medial) but the *position* within the MG of the homologous region may vary across species.

The *ventral* division contains neurones with dendritic fields aligned parallel to each other and to the incoming fibres of the brachium of the inferior colliculus. Neurones are sharply tuned to sound frequency and their CFs form an orderly tonotopic array (Calford and Webster, 1981; Imig and Morel, 1985). Fibres terminating in the ventral division are derived principally from the CNIC, and it is appropriate to consider the ventral division of the MG as constituting the core relay of the auditory thalamus, with efferent axons projecting to the primary auditory cortex.

The *dorsal* division is populated by neurones with radiate dendritic fields that do not form laminar arrays. They are more broadly tuned to tonal stimuli and often have poorly defined CFs. They receive afferents from sources outside the CNIC, including the dorsal cortex and nucleus sagulum. Efferents pass not only to the auditory cortex, but also to other non-specific forebrain regions.

The *medial* division has a variety of afferent sources, including the external nucleus of the inferior colliculus and somatosensory areas. Some of its neurones are very sensitive and sharply tuned to tonal stimuli; others are more broadly tuned. Efferents of the medial division terminate widely in all auditory cortical areas and also in parts of the somatosensory cortex.

The lateral posterior nucleus, *Po*, is contiguous rostrally with the ventral and medial divisions of the MG. It contains neurones with sharply tuned responses (Phillips and Irvine, 1979a) and receives input from the CNIC. Efferents from Po ascend to terminate principally in the anterior auditory field and to a lesser extent the primary auditory cortex.

Thus, at the threshold to the auditory cortex, several major streams of information flow can be recognized, although this description has concentrated on that issuing from the CNIC. This core system is undoubtedly the largest component of the ascending auditory pathway, but it should be remembered that the external nucleus and dorsal cortex of the inferior colliculus also provide information destined for the auditory cortex. Further observations will be made about the physiological properties of neurones in these tecto-thalamo-cortical pathways later in this book. Let us now turn our attention to the auditory cortex and examine the physiological basis for its subdivision.

4 The representation of the cochlea in the cerebral cortex

The evolution of the auditory cortex of mammals relates to the characteristic organization of all sensory cortical areas into a central koniocortex, surrounded by belts of cortex related to the limbic, olfactory or insular regions (Sanides, 1975). During evolution, a dense focus of thalamic input must have arisen within the para-insular temporal cortex; this new zone became the *primary* auditory cortex. During later development, the linking para-insular cortex lagged behind and formed less precise relationships with the MG; these became secondary auditory areas. In all mammalian species so far studied there is at least one primary representation of the cochlea in the cerebral cortex. By 'representation' I mean (i) that sound-evoked electrical activity originating from one part of the cochlea is localized in one or more discrete regions of the cerebral cortex and (ii) that there is a systematic spatial relationship or mapping between places of activity along the cochlea and across the cerebral cortex. This representation is referred to as 'cochleotopic' or 'tonotopic' organization.

A cortical representation of the cochlear frequency organization is arranged so that the frequency represented at a single cochlear locus is mapped on to a three-dimensional cortical slab. One assumption underlying this general description is that the frequency represented by this cortical slab is the most effective frequency for all of the neurones occupying that cortical volume. Although neurones closely situated in an individual electrode penetration are reported to have very similar characteristic frequencies (CF; the frequency at which the threshold of a unit is lowest) (Merzenich *et al.*, 1975; Phillips and Irvine, 1979b), this assumption has never been fully tested because most experiments have been carried out in anaesthetized animals and anaesthesia appears to depress the excitability of cells in both the superficial and deep cortical layers. Evans and Whitfield (1964) note that in some penetrations into auditory cortex of unanaesthetized cats, units with widely differing CFs may be recorded in close proximity. Similar observations were made by Hind (1960) for anaesthetized cats, although he indicates that the monaural or binaural nature of the stimulus may have been a determinant. Furthermore, in unanaesthetized animals many auditory cortical units do not respond to tones or respond only sluggishly (Evans and Whitfield, 1964; Funkenstein and Winter, 1973). Unfortunately the technique of recording from unanaesthetized, awake animals usually precludes the possibility of relating unit response properties to a given depth or given location, but it would seem probable that not all units in a given slab of auditory cortex respond to tones.

We will return to the question of the organization of the auditory cortex as a function of depth in Chapter 6, but at this point it is important to remember that maps of cochlear representation in the cerebral cortex are probably maps of activity in the middle cortical layers, which receive the bulk of thalamic afferent input. The measure of frequency selectivity usually employed in mapping studies is the CF and it is known that the frequency ranges to which units respond expand at intensities above threshold. For some units the best excitatory frequency can shift from the CF to higher or lower values at higher intensities. As a consequence, maps of cochlear representation are not necessarily fixed for all intensities of sound.

AUDITORY CORTICAL AREAS IN PRIMATES

Woolsey has recently described the experiments carried out with Walzl in the 1940s to demonstrate cochleotopic organization in the auditory cortex of the macaque monkey (Woolsey and Walzl, 1982). These experiments and those of Licklider and Kryter (1942) laid the foundation for later studies using microelectrode mapping techniques that allow for fine-grain analysis of cochlear representation. These have been carried out on the macaque monkey (Merzenich and Brugge, 1973), owl monkey (Imig *et al.*, 1977), galago (Brugge, 1982) and marmoset (Aitkin *et al.*, 1986b).

The auditory cortex of the macaque monkey is located within the lateral fissure, on the superior temporal plane (Figure 4.1, upper). Microelectrode penetrations made normal to the superior temporal plane (exposed by ablation of overlying parietal cortex) reveal an area where units respond strongly and at short latency to tone bursts. The region surrounded by filled circles in the lower part of Figure 4.1 is considered to be the primary auditory area (AI) on the basis of responses to tones and on the characteristic granular cytoarchitecture described in Chapter 1. Merzenich and Brugge (1973) note that in the AI of macaques, the loci of neurones with similar CFs can be joined by an imaginary line to construct 'isofrequency lines', with lines for low CFs found rostrolaterally and those for high CFs caudomedially. This systematic

EXPT. 70–200

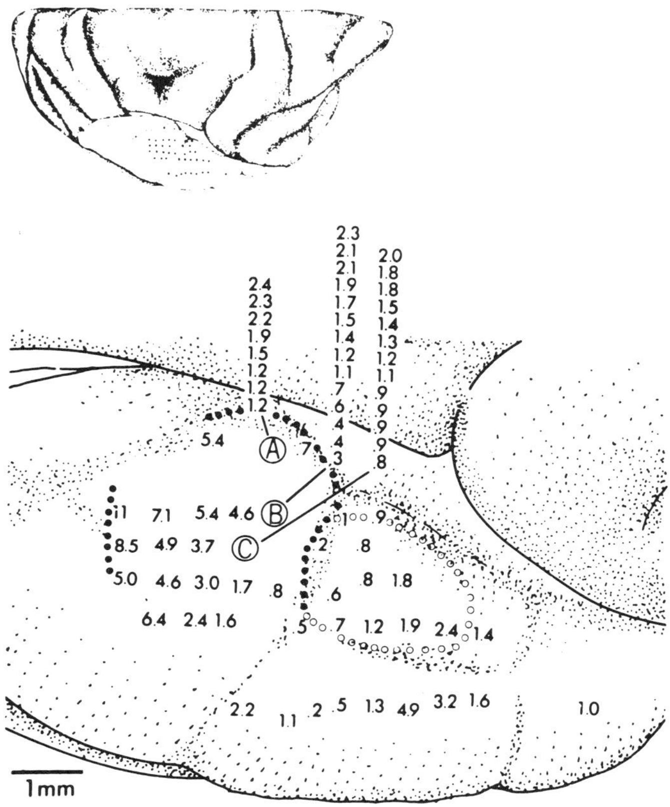

Figure 4.1 Surface map showing the distribution of CFs of 42 penetrations into the primary auditory cortex of a macaque monkey, and fields rostral and lateral to it. Approximate boundaries of AI, except laterally, are indicated by the dotted line. Line of small open circles encloses the small rostrolateral field (RL). Penetrations labelled A, B and C pass down the bank of an elevation on the superior temporal plane. Columns of numbers with these points are CFs of neurones encountered at regular intervals within these penetrations. From Merzenich and Brugge (1973).

shift in CF is also seen when microelectrode penetrations are oriented orthogonal to the direction of isofrequency lines and run through the granular layer tangential to the cortical surface. Sequences of CFs from cells encountered along three tangential electrode penetrations, marked A, B and C, are seen in Figure 4.1, lower.

There are other auditory cortical fields adjacent to the AI of macaques, although the organizations of these fields are less clear. A rostrolateral field (Figure 4.1, lower) has a similar cytoarchitecture to the AI and shares a common low-frequency border, but appears to have a restricted high-frequency representation. Lateral to the rostrolateral field, and close to the lip of the lateral fissure, CFs may be recorded which appear not to belong in sequence to either the AI or the rostrolateral field (lowermost row of CFs, Figure 4.1, lower). Finally a caudomedial field with distinctive cytoarchitecture contains broadly tuned units, the CFs of which are difficult to specify.

The owl monkey auditory cortex contains five possible fields, the AI and fields rostral, anterolateral, posterolateral and caudomedial to the AI (Imig *et al.*, 1977). As in the macaque, low frequencies are located laterally and rostrally within the AI, and high frequencies caudally and medially. Isofrequency lines are, however, often irregular in shape and the primary field cannot be described as a linear matrix of isofrequency contours. There is a sudden change in CF rostral to AI, indicating a border with the rostral field. Constituent unit CFs rise steadily from this border to the rostral limit of the rostral field. This contrasts with the mirror image symmetry of frequency changes about the border between the AI and rostral field of macaques. As in the latter species, evidence for other complete cochlear representations is rather meagre in owl monkeys.

The auditory cortex of marmosets contains one major field, AI, surrounded by zones in which the organizations of CFs are not clear (Aitkin *et al.*, 1986b). The zone of granular cortex varies in the extent to which it is found on the ventral bank of the lateral sulcus and on the adjacent temporal gyrus. Correspondingly, in some marmosets, area AI may occur largely on the surface and in others has a significant segment on the ventral bank of the lateral sulcus. The map of Figure 4.2 (upper) reveals a similar orientation of the AI to the two primates discussed above, spreading from below 1 kHz rostroventrally to 30 kHz caudodorsally. The array of CFs is orderly but has a complex shape and isofrequency lines are not parallel. The CF changes monotonically as a function of depth in the ventral bank of the lateral sulcus. This is shown graphically in Figure 4.2 (lower) for six penetrations in two experiments when penetration depth is plotted against CF. After a period of constancy, CF changes to lower values with caudal penetrations or to higher values with rostral tracks. The cochlear representation in the ventral bank of the lateral sulcus is complex and extensive. It is not clear if it belongs to the AI or is a separate strongly responsive field, perhaps analogous to the rostral fields of the macaque and owl monkey.

Electrophysiological evidence for the existence of secondary fields is

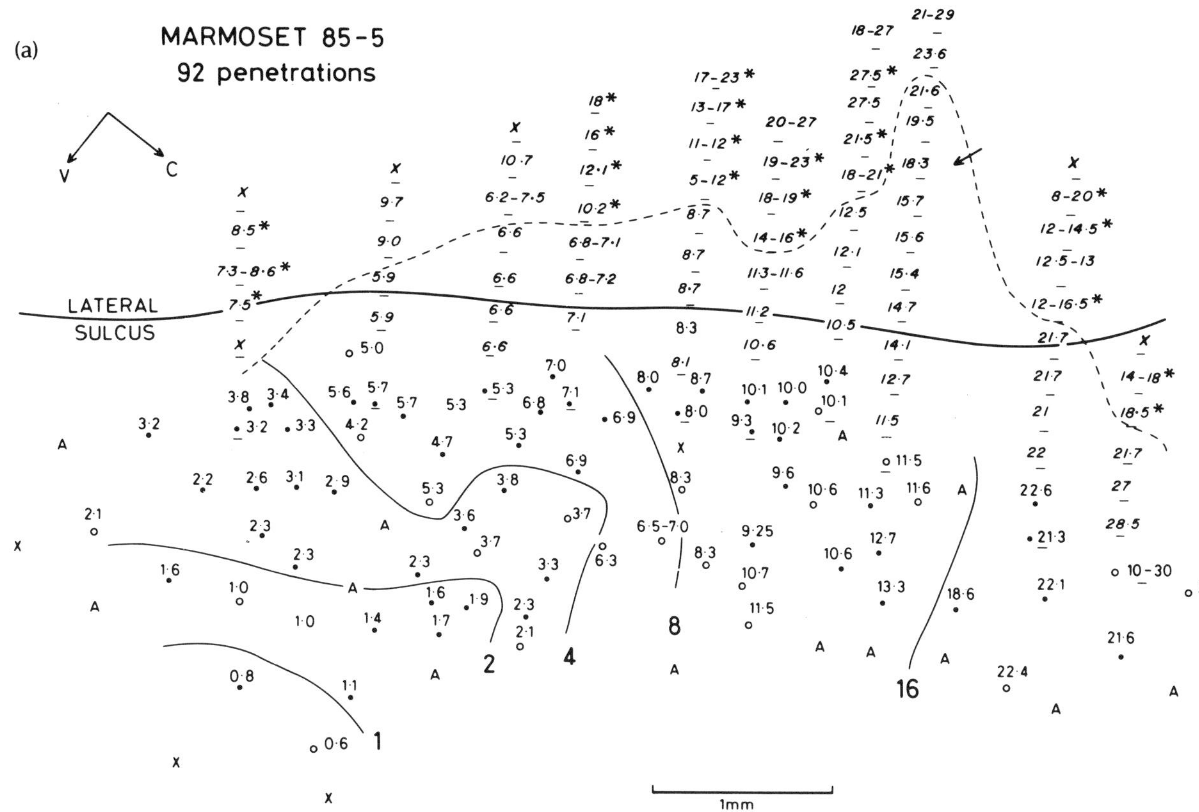

Figure 4.2 (a) The distribution of CFs of neurones and neurone clusters across the acoustically responsive cortex in a marmoset. Filled circles indicate strongly responsive cortical loci and open circles weakly responsive points; sites marked with 'A' responded to auditory stimuli, but no CF could be defined; those indicated by 'X' were unresponsive to sound. CFs in italics are those observed in penetrations down the ventral bank of the lateral sulcus (orthogonal to the gyral surface), with asterisks indicating weak points. The distances between these points can be calculated from the calibration on the figure and the distance between dashes beneath each sulcal bank CF. The dashed line extending above the lateral sulcus shows the approximate boundary of area AI in the lateral sulcal bank. Isofrequency contours in octave steps have been estimated and labelled 1, 2, 4, 8 and 16 kHz. C, caudal, V, ventral.

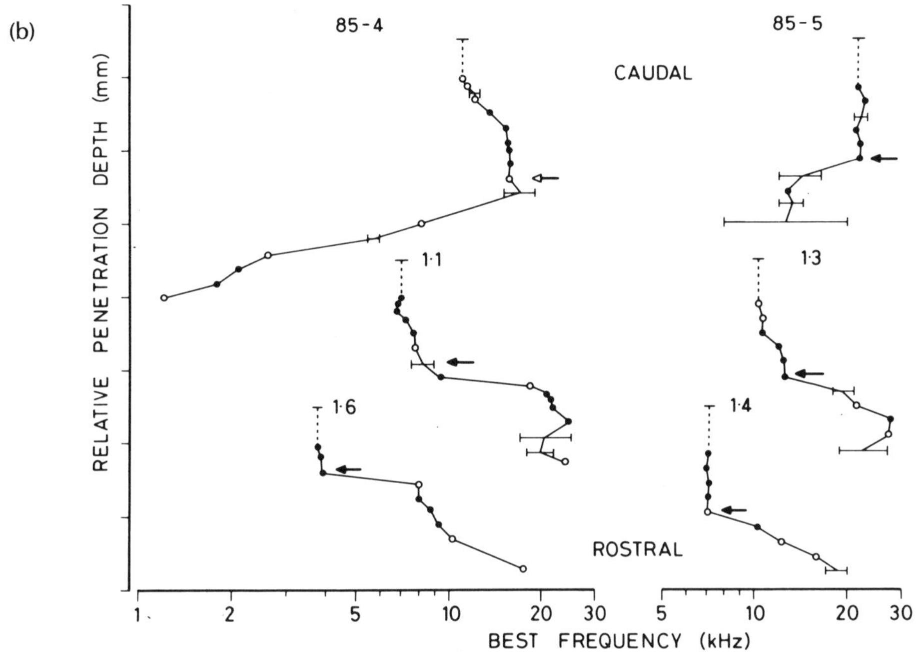

Figure 4.2 (b) The sulcal bank penetrations in each of two marmosets with a very caudal penetration (top) separated by a middle penetration from a track 2.7 mm rostrally. The numbers attached to the middle and lower penetration sequences indicate the distances of those tracks from the caudal track above, in millimetres. The penetrations share a common relative penetration depth axis (ordinate) that is pertinent for points within a track but not between tracks. The relationship between pial entry point and the first CF measurement is indicated with a broken line. Broadly tuned clusters are shown with a line connecting upper- and lower-frequency limits. Filled and open circles indicate strong and weak responses. The arrows indicate discontinuities in CF sequences, and show the extent of the AI, as measured physiologically, down the sulcal bank. From Aitkin *et al.* (1986b).

weak, aside from the discontinuities illustrated for the ventral bank of the lateral sulcus mentioned above. Changes in the CF are noted rostrally and rostroventrally (Aitkin *et al.*, 1986b) but the observations are not extensive enough to allow for parcellation into fields. There is, however, more persuasive evidence for secondary fields on the basis of anterograde tracer injections into the AI (Aitkin *et al.*, 1988a); these will be discussed more extensively in Chapter 5.

The cytoarchitecture of the auditory cortex of humans has been described in some detail (Galaburda and Sanides, 1980) and its location, but not its tonotopic organization, identified electrophysiologically (Celesia, 1976). Isotope measurements of regional cerebral blood flow have confirmed that an area of cortex corresponding to that studied histologically and electrophysiologically is activated by pure tones (Roland *et al.*, 1981). Positron emission tomography (see Chapter 2) has been applied in a study by Lauter *et al.* (1985) to show that the regions of maximum cerebral blood flow in this area differ consistently with low (0.5 kHz) and high (4 kHz) frequency tone stimulation. The greatest activity change evoked by 0.5 kHz was in an area lateral and anterior to that activated by 4 kHz. If these areas are parts of the AI it would appear that the primary cochlear representation has a similar orientation to those of the subhuman primates. Measurements of magnetic fields evoked by pure tones, made using the SQUID (see Chapter 2), have indicated a logarithmic relationship between distance across a plane 2–3 cm below the scalp, and frequency giving greatest neuromagnetic response (Romani *et al.*, 1982). Such a relationship is similar to that seen in other mammals, but it is difficult to relate this functional measurement to anatomical descriptions of the auditory cortex of humans. These important studies need to be extended but this will depend upon dramatic reductions in the cost of using these non-invasive techniques.

AUDITORY FIELDS IN NON-PRIMATE MAMMALS

The cat has generally been accepted as a model for the organization of the auditory cortex, but this may be more of historic than functional significance. The common availability of cats, their ease of maintenance, their robust nature when anaesthetized, and acoustic sensitivity have dictated their use when possible by students of neurophysiology for the last 50 years. This has resulted in a large database that is constantly being added to, and the natural tendency for neurophysiologists to see the cat as, in some way, archetypal of mammals. It is certainly true that the behavioural audiograms of cats show very broad sensitivity, from 60 Hz to 60 kHz (Heffner and Heffner, 1985; Neff and Hind, 1955), and that all aspects of the brain stem auditory system (e.g. the superior olivary complex) are well developed in cats.

Auditory fields have been mapped in cats since the studies of Woolsey and Walzl (1942) employing electrical stimulation of the osseous spiral lamina. They identified complete cochleotopic organizations in the AI (low-frequency caudal, high-frequency rostral) and secondary areas AII (reversed tonotopic organization to AI) and Ep (high-frequency dorsal, low-frequency ventral). An area rostral to the AI was shown to occupy the suprasylvian sulcal bank and to spread out dorsal to the AI on the ectosylvian gyrus (high frequency) and ventral on the anterior ectosylvian gyrus (low frequency).

The microelectrode was subsequently employed to map the AI with tonal stimuli, by Hind (1953), Evans *et al.* (1965) and Goldstein *et al.* (1970). They used differing acoustic stimuli, varying techniques of immobilization and different definitions for best excitatory frequency. All generated composite maps on which the preferred frequencies recorded in many experiments were superimposed on an outline of the auditory cortex defined by the sulcal landmarks, using Woolsey's (1960) map. All studies showed a tendency for the preferred excitatory frequency to shift from low, caudal, to high, rostral in the AI in accordance with the study of Woolsey and Walzl (1942). All studies, however, showed many inconsistencies with the classical model of the AI.

Resolution of some of these inconsistencies was obtained in the study by Merzenich *et al.* (1975). They presented contralateral pure tones in barbiturate-anaesthetized cats and recorded the CFs of clusters of units isolated by microelectrodes from the middle cortical layers. They found that the CFs obtained with multiple penetrations in individual animals shifted systematically from low caudally to high rostrally. The use of individual maps, of anaesthetized preparations and of cluster recordings of CFs allows the basic pattern of organization to be revealed.

Secondary fields have also been redefined with microelectrode recording (Figure 4.3). Knight (1977) and Reale and Imig (1980) showed that an anterior field shared the high-CF border with the AI and its tonotopic axis curved rostrally along the anterior ectosylvian gyrus. This field resembles the rostral fields of macaques and owl monkeys in the strength of the responses that can be evoked within it. It probably incorporates part of the suprasylvian fringe area of Woolsey (1960).

Two tonotopic fields occupy the region posterior and posteroventral to the AI, largely within the caudal bank of the posterior ectosylvian sulcus (Figure 4.3). These together are probably identical to the posterior ectosylvian area of Woolsey (1960). Neurones in the second auditory field (AII) and ventral, temporal and dorsoposterior areas (Figure 4.3), although responsive to acoustic stimulation, are often poorly tuned, so

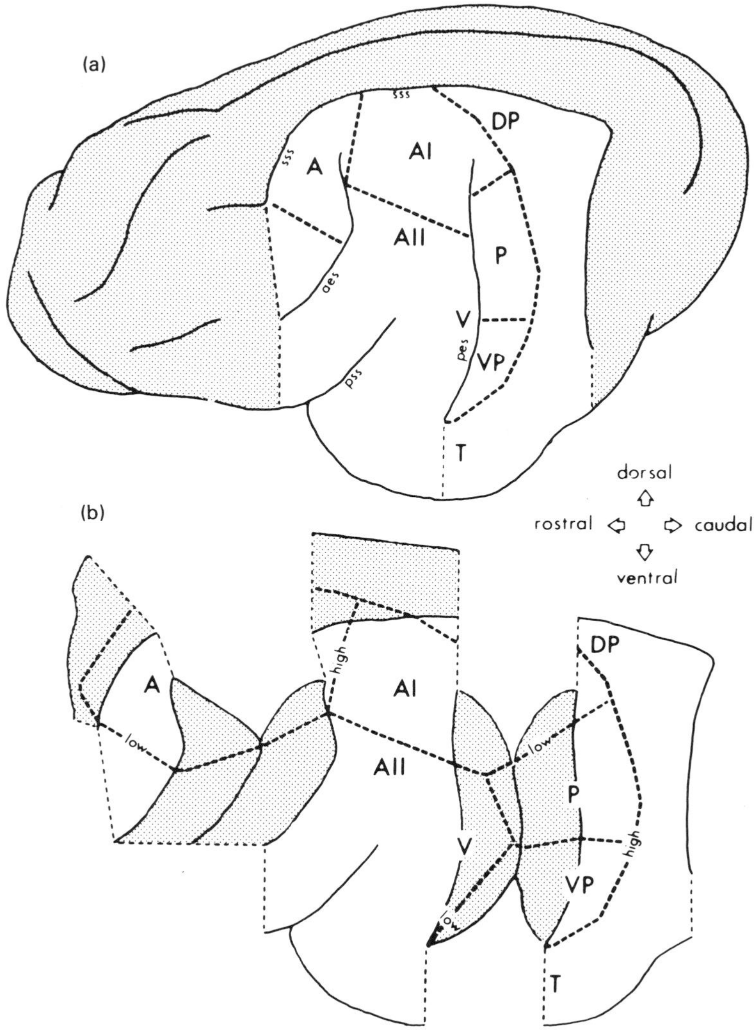

Figure 4.3 Auditory cortical fields in the cat. (a) A lateral view of the left cerebral hemisphere showing positions of fields. (b) Unfolded cortical surface forming the gyral surfaces and sulcal banks in the unshaded regions in (a). Cortical surfaces forming sulcal banks are shaded while those forming gyral surfaces are not. Heavy interrupted lines delimit four tonotopically organized fields: anterior, A; primary, AI; posterior, P; ventroposterior, VP. The locations of lowest and highest CFs in these fields are indicated in (b) by 'low' and 'high,' respectively. Surrounding these four fields is a belt of cortex containing neurones responsive to acoustic stimulation: V, ventral; T, temporal; DP, dorsoposterior areas. sss, suprasylvian sulcus; aes, pes, anterior and posterior ectosylvian sulci; pss, pseudosylvian sulcus. From Imig and Reale (1980).

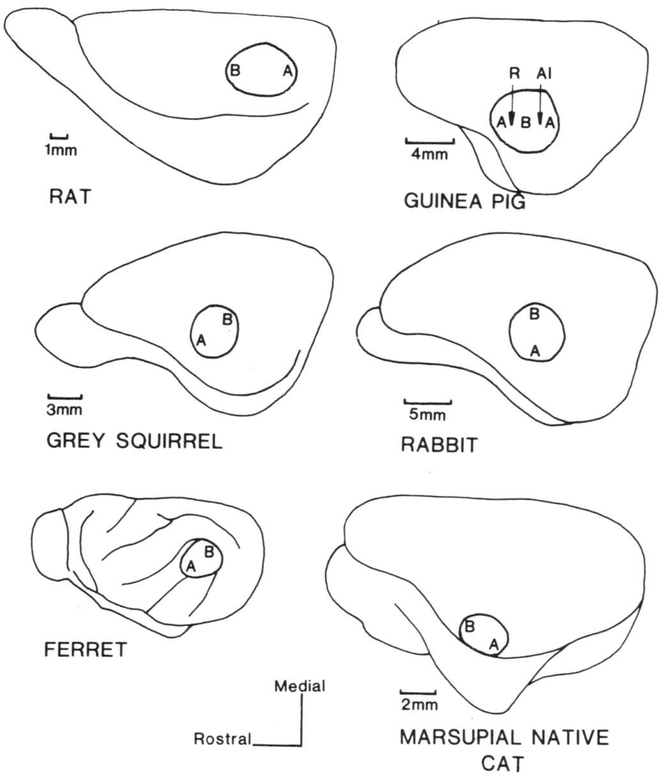

Figure 4.4 Left cerebral hemisphere of six mammals to show approximate locations of primary auditory area, AI (plus the rostral field, R, for the guinea pig) and the cochleotopic axis (base, B, to apex, A).

that tonotopic maps have not been obtained for all regions. Schreiner and Cynader (1984) conclude that there is a high variability in organization in AII from one cat to the next. They propose that AII 'contains general organizational gradients across the entire field overlayed by strong local gradients' (p. 1285).

For most non-primates other than cats, the AI is the only field to have been completely mapped, although some authors note the presence of discontinuities in frequency progressions at the edge of the AI. Aside from echo-locating bats, which will be examined shortly, detailed maps have been presented for rabbits (McMullen and Glaser, 1982), grey squirrels (Merzenich *et al.*, 1976), rats (Sally and Kelly, 1988), ferrets (Kelly *et al.*, 1986), guinea pigs (Hellweg *et al.*, 1977) and marsupial native cats (Aitkin *et al.*, 1986a). These are summarized in Figure 4.4.

In all six species the AI is located in the temporal lobe. In the marsupial native cat it lies adjacent to the rhinal sulcus but in the eutherian mammals there is a zone of neocortex between the rhinal sulcus and the AI. If this region is homologous to that in the cat, it is putatively auditory in function. The tonotopic axis – the imaginary line connecting the representations of the basal-most (B) and apical-most (A) segments of the cochlea – varies from species to species (Figure 4.4). In rats, guinea pigs and native cats it is nearly parallel to the rostro-caudal axis. Note in guinea pigs the presence of two fields containing strongly responsive units, the rostral field and the AI, both aligned to the rostro-caudal axis. The tonotopic axis is tilted and reversed in squirrels, ferrets and rabbits – in the latter species it is almost orthogonal to the rostro-caudal axis. Preliminary observations on gerbils (Steffen *et al.*, 1988) indicate several fields in this species, and the AI shows a tonotopy from apex dorso-caudally to base rostrally, as in rats.

Suga and co-workers have provided, for the mustache bat, the most detailed accounts of the auditory cortical representations of frequency, amplitude and biosonar information for any species. These echo-locating bats emit a long constant-frequency signal terminated by short frequency-modulated components. In a mapping study, Suga and Jen (1976) described a disproportionate tonotopic representation in the auditory cortex of the mustache bat (Figure 4.5). High frequencies (65–100 kHz) are represented rostrally and low frequencies (20–50 kHz) are found caudally in a large area of the temporal neocortex (Figure 4.5A). In the centre of this region neurones whose CFs lie between 60 and 63 kHz (corresponding to the constant-frequency signal) occupy approximately one third of the rostro-caudal auditory cortical axis. Isofrequency contours are arranged in a circular fashion in this zone. It is clear from the summary map of Figure 4.5b that other regions with orderly progressions of characteristic frequency are found dorsal and ventral to the putative AI, shown between the dotted lines. Suga and co-workers (e.g. Suga, 1982) have described these regions as including centres for the processing of combinations of frequency-modulated stimuli, and for comparing the frequencies of constant-frequency emissions and their Doppler-shifted echoes.

It should be noted that this complex organization of the auditory cortex in the mustache bat is not necessarily typical for all bats. Suga (1965) and Wong and Shannon (1988) studied the auditory cortex of the bat *Myotis lucifugus*, also an echo-locator, and found a fairly uniform representation of cochlear place as a function of location in the AI. Wong and Shannon (1988) found no evidence for a disproportionate representation of any bandwidth but argued for the existence of an auditory cortical zone that contained neurones with different preferred

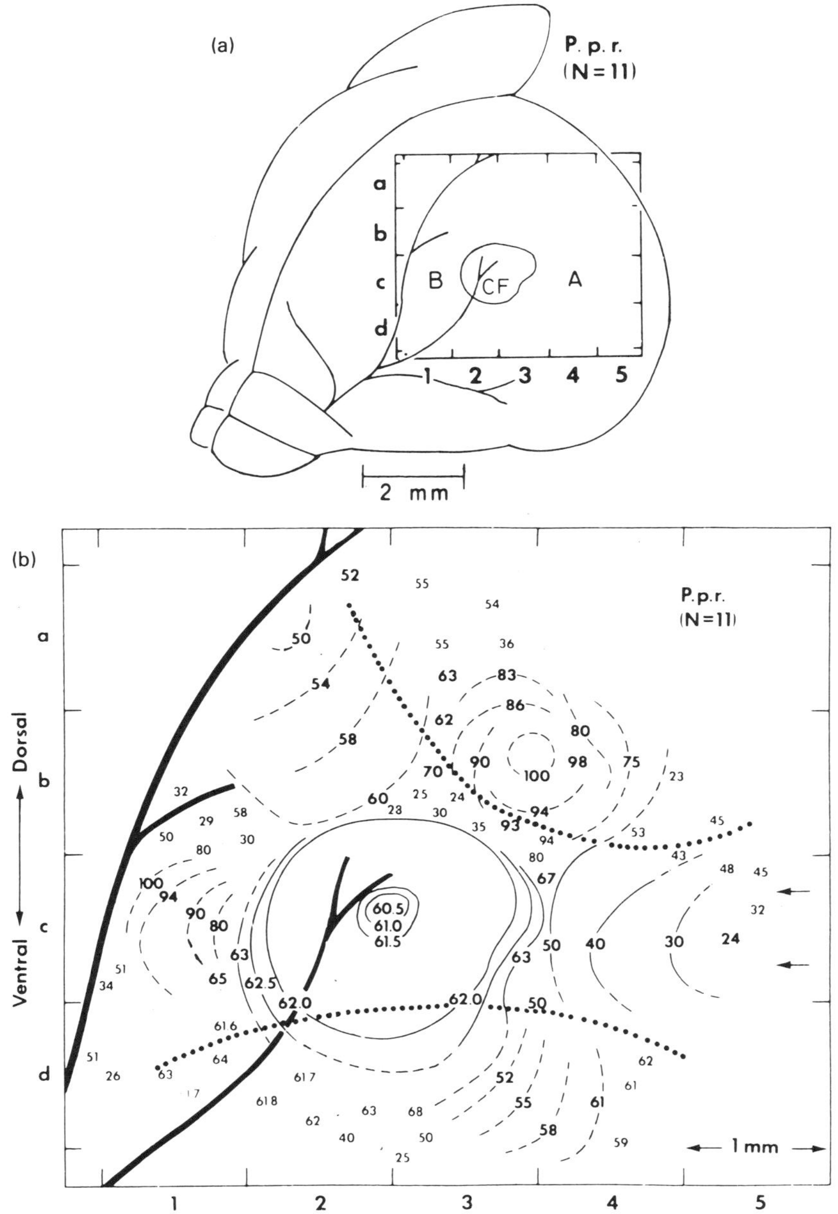

Figure 4.5 (a) Dorsolateral view of cerebrum of mustache bat. Rectangle includes auditory areas. A, apex; B, base; CF, constant frequency area. (b) Distribution of characteristic frequencies in the rectangle shown in (a). Area between dotted lines is the AI. Solid lines, clear isofrequency contours; dashed lines, frequency organization vague. Composite map from 11 bats. From Suga and Jen (1976). Copyright 1976 by the AAAS.

interaural time delays, important as a basis for the coding of target distance.

The tonotopic map in the auditory cortex of the mustache bat varies in detail from one specimen to the next, according to the particular value of the constant-frequency call: the higher the call, the higher the frequencies represented (Suga *et al.*, 1987). There is also some evidence for sexual dimorphism in the frequency of the constant-frequency component, with that of the female mustache bat higher than that of the male (Suga *et al.*, 1987). Suga and co-workers write of this variation as a 'personalization . . . suited for the detection of wing beats of insects . . . [and] . . . also for the reduction of the masking effect on the echolocation of sounds produced by conspecifics' (Suga *et al.*, 1987, p. 652). Further analysis of this important series of studies of the auditory cortex of a remarkable mammal will be given later in this book. It is of interest to note, however, that the mustache bat is the only species yet examined in which the cortical map shows a highly disproportionate representation of frequency.

How systematic and linear are the maps of characteristic frequency in the AI? Early studies such as those of Tunturi (1952) suggest a rigid geometry in the relationship between cortical place and preferred frequency. In their studies of the AI of macaque monkeys, Merzenich and Brugge (1973) introduced the idea that iso-CF points could be connected by straight lines on a cortical map. This procedure was elaborated in the study of Merzenich *et al.* (1975) who wrote (p. 233) 'It was also evident in all cats that any given frequency band was represented across a belt of cortex, aligned on a nearly straight mediolateral axis, and extending across the width of the primary field.'

Close inspection of the maps of the study of Merzenich *et al.* (1975), however, suggest that the straight lines connecting the same CFs were dictated by the experimenter, rather than the maps. The illustrated lines were rarely, if ever, shown to connect identical measured CFs and often much extrapolation or interpolation was used. It is tempting to consider the frequency organization of AI as completely orderly and linear in all species, but this generalization must be viewed with caution given the way isofrequency lines were drawn in cats, the variation in the locations of maps *re* the sulcal landmarks shown by Merzenich *et al.* (1975), and the likelihood of specimen-to-specimen variability in the internal organization of the maps, suggested, for example, in the work of Suga *et al.* (1987). Published maps for marmosets (e.g. Figure 4.2) give little indication of a linear matrix relating CF to cortical position, even though there is a systematic relationship between CF and cochlear place along any imaginary rostro-caudal line through the AI.

Is the representation of frequency continuous, or are there quantal steps in frequency between adjacent units? Merzenich *et al.* (1975) showed for the AI of cats that the relationship between cochlear place (derived from a conversion from frequency to cochlear place made by Greenwood (1961)) and cortical distance was smooth and that basal segments occupied more cortical distance than did apical segments. Large step changes in CF were rarely encountered as the microelectrode was moved across the cortical surface. A similar smooth relationship has been shown for marmosets, in which extremely dense mapping studies have been carried out (Aitkin *et al.*, 1986b). For most mammals the basal segments occupy more space than do apical cochlear segments in the AI, but this is a smooth, rather than a disproportionate, representation.

FUNCTIONAL ASPECTS OF TONOTOPIC ORGANIZATION

As previously mentioned, the non-invasive PET technique has indicated that the AI of humans is likely to be tonotopically organized. Such a technique assesses functional activity in terms of changes in blood flow to an area. Functional activity may be measured more precisely in animals by using the 2DG technique, but this has so far been less convincing in the auditory cortex of mammals. Steffen *et al.* (1988) have had some success with ^{14}C-fluoro–2-deoxyglucose in showing discrete banding in the auditory cortex of gerbils that corresponds to pure tone presentation. However, when conventional 2DG was used, only faint effects were observed by Ryan *et al.* (1982) for alert gerbils, and Hungerbühler *et al.* (1981) reported that pure tone stimulation did not alter glucose utilization in the auditory cortex of either anaesthetized or awake cats. In contrast, both narrow-band and wide-band noise stimuli produce activity in this region in both cats (Hungerbühler *et al.*, 1981) and gerbils (Sharp *et al.*, 1981). A similar negative result in relation to pure tone stimulation was observed earlier by Webster *et al.* (1978).

Given the strong uptake shown in lower auditory centres, this poor utilization by the auditory cortex is puzzling. It is possible that the low neuronal firing rates evoked in many cortical neurones by tonal stimuli might be associated with a low, non-discriminable increase in glucose utilization (e.g. Theurich *et al.*, 1984). It is possible that the significance of the stimulus may have a progressively greater importance at the cortical level. Even at the inferior colliculus, classical conditioning procedures can enhance the uptake of 2DG (Gonzalez-Lima and Scheich, 1984). In this regard, clear tonotopic 2DG banding has been described in a ‘simpler’ auditory cortex, the neostriatum of the chick (Scheich, 1983). It would not appear that the complexity *per se* of the mammalian cerebral cortex renders it unsuitable for this technique, since clear banding occurs in the visual (e.g. Humphrey *et al.*, 1980) and somatosensory (e.g. Hand *et al.*, 1977) cortex of mammals with appropriate stimulation.

Is tonotopic organization necessary for frequency discrimination? If so, it might be expected that discrete lesions of the AI would result in a restricted loss of frequency discrimination. Such discrete lesions are never likely to occur in humans but can be artificially produced in experimental animals.

Studies by Meyer and Woolsey (1952) (cited in Rose and Woolsey, 1958) and by Neff and co-workers (Butler *et al.*, 1957; Goldberg *et al.*, 1958; Neff and Diamond, 1958) have shown that cats are able to make frequency discriminations after complete bilateral lesions of AI. However, frequency detection is impaired if frequency is tied to other attributes of an auditory stimulus such as its location in space. Thus, cats can be behaviourally deaf to a certain frequency, following a lesion to the AI, if the sound is presented from a certain region in space (Jenkins and Merzenich, 1984). It may not therefore be appropriate to study frequency discrimination without relating the frequency to location in space, or to other parameters such as intensity, temporal pattern, duration and modulation. It must be concluded at this stage that the evidence from lesion studies in animals and humans has little relevance to the issue of the function of tonotopic organization of the AI.

5 Connections of the auditory cortex

It is the purpose of this chapter to document the main pathways afferent and efferent to the auditory cortex. Some of the observations to be made have functional or structural ramifications that cannot be considered at this stage, but will be returned to later in the book. Such issues include questions of the laminar termination of fibres within the auditory cortex, and cells of origin of efferent pathways. Much recent work combining pathway-tracing techniques with electrophysiological mapping has been carried out in cats, but information about gross connectivity in other species will be linked to studies in cats where possible.

ASCENDING CONNECTIONS

Fibres leaving the MG ascend in the acoustic radiations to the auditory cortex, where those from the ventral division terminate in the middle and deepest cortical layers. The gross connectional relationships between the MG and auditory cortex, as summarized in Chapter 3, have been unravelled from studies of retrograde degeneration of neurones in the MG following lesions of the auditory cortex, from observations of anterograde degeneration in the cortex after lesions of the MG, and from tracing studies using relatively large volumes of tracer (see Winer, 1985a, for references).

The details of these connections and their topographies have been provided by experiments in which the tonotopic organization of the auditory cortex has been mapped in a given animal and then small injections of retrograde or anterograde tracers have been made at selected, mapped sites (Aitkin *et al.*, 1988b; Andersen *et al.*, 1980a,b; Luethke *et al.*, 1988; Middlebrooks and Zook, 1983; Morel and Imig, 1987).

Morel and Imig (1987) consider the connections between different subdivisions of the auditory thalamus and cortex in cats in terms of major and minor projections. Thus, for example, the AI receives a major input (in terms of numbers of labelled cells) from the ventral division and minor inputs from the medial division and Po. The anterior field, on the other hand, receives major inputs from the medial MG and Po,

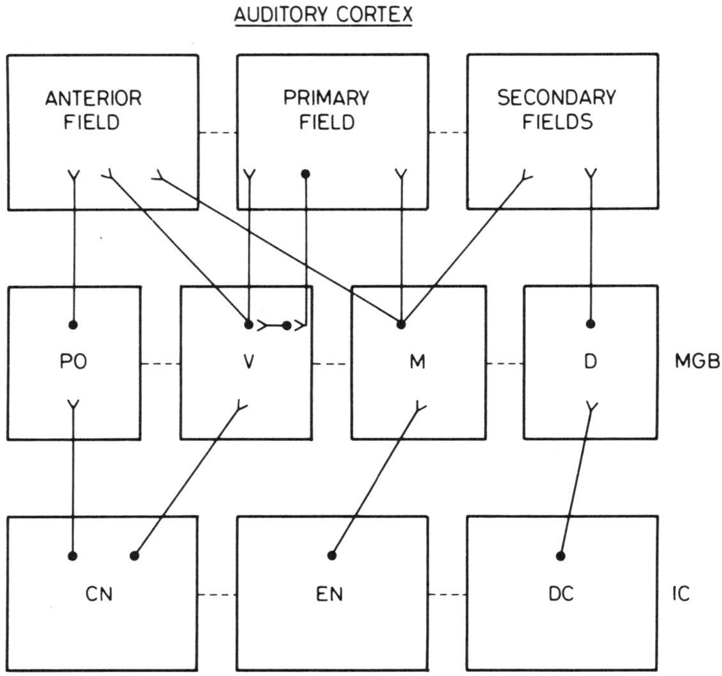

Figure 5.1 Block diagram showing the principal connections of the upper auditory pathway of cats. Medial geniculate body (MGB), ventral (V), medial (M) and dorsal (D) divisions, posterior nuclear group of the thalamus (PO), inferior colliculus (IC), central nucleus (CN), external nucleus (EN) and dorsal cortex (DC).

and a minor projection from the ventral division. Fields P and VP receive major inputs from different parts of the ventral division and different parts of the dorsal division. Morel and Imig suggest that different, segregated regions of the ventral division and Po contain many neurones projecting to one field and only sparse numbers projecting to others. Some of these features are summarized in Figure 5.1.

Mapping and tracing studies show that connections between the ventral division and the AI are both precise and reciprocal. Merzenich *et al.* (1982) interpret their results in cats as showing that discrete regions of the AI receive input from complex, folded bands of neurones in the ventral division of the MG. The locations of the bands in the MG shift systematically as a function of the site of the tracer injection in the AI. If tracer injections occupy a substantial part of an isofrequency slab, the band of labelled cells in the MG is unbroken, but with small injections, gaps occur in the bands of labelled cells. Middlebrooks and Zook (1983) relate these discontinuities to binaural bands in the AI, a topic I shall return to later.

The importance of corticofugal projections to the MG was first revealed in anterograde degeneration studies following lesions to the auditory cortex (Diamond *et al.*, 1969). All medial geniculate divisions receive descending connections, but tracer injections have revealed that, in the ventral division, these descending projections are not only substantial in numbers but are also highly topographic – i.e. that there is a precise reciprocal connection between the AI and the ventral division of the MG. Morest (1975) has shown that the corticogeniculate fibres synapse not on the principal relay cells of the ventral division but on the cell bodies and proximal dendrites of interneurones, where they would appear to make excitatory synapses. The synaptic endings of the interneurones have cytological features shared by inhibitory synapses, so that it is possible that the activation of corticofugal fibres may have inhibitory effects in the MG. Electrophysiological studies confirm this possibility (Aitkin and Dunlop, 1969; Watanabe *et al.*, 1966).

These findings suggest that activity at a site in the auditory cortex must be seen as the result of intrinsic cortical synaptic circuitry interacting with continuous feedback and feedforward activity of thalamocortical loops. Such an idea is not a novel one but was enunciated almost forty years ago (Chang, 1950).

Ascending input to the auditory cortex of cats is also derived from non-auditory sources. Injections of retrograde tracers into the AI, A or P fields label neurones in the locus coeruleus, lateral hypothalamus and tegmentum, whereas the AII receives extrathalamic input mainly from the locus coeruleus (Rouiller *et al.*, 1989). The significance of such projections remains to be determined, but possibilities include roles in plasticity and arousal.

INTERHEMISPHERIC CONNECTIONS

In cats, all subdivisions of the auditory cortex give and receive projections from the auditory cortex in the opposite hemisphere (Diamond *et al.*, 1968a; Kawamura, 1973; Imig and Reale, 1980). The primary cortex projects to the AI and AII of the opposite side and to the tonotopically organized fields located anteriorly and posteriorly. The AII projects to the opposite AII, while anteriorly and posteriorly situated fields project to their homotypic areas and to the AI in the opposite hemisphere.

In tracing experiments in cats in which the auditory cortex was electrophysiologically mapped bilaterally and large injections of tritiated amino acids were made into one auditory cortex, callosal fibres were shown to terminate most densely in regions of the AI containing binaurally excitatory neurones (Imig and Brugge, 1978). However, removal of callosal input by sectioning the callosum did not eliminate the binaural interactions observed in cortical neurones, so that callosal input is not responsible for these interactions, although it may modulate them (Imig and Brugge, 1978).

Amongst primates, interhemispheric connections have been studied in rhesus monkeys (Pandya *et al.*, 1969; Karol and Pandya, 1971), owl monkeys (Fitzpatrick and Imig, 1980) and marmosets (Aitkin *et al.*, 1988a). The auditory cortex of rhesus monkeys, located using cytoarchitectural criteria, projects only to the medial part of the contralateral auditory cortex. Pandya *et al.* (1969) speculated that, in keeping with visual callosal projections, the auditory connections might carry information about sounds located in the midline and that spatial mapping might therefore be from lateral to medial across the auditory cortex, orthogonal to the tonotopic axis. Although this suggestion remains to be tested in subhuman primates, studies of the minimum audible angle in cats after callosal section show no obvious deficits in localization thresholds (Moore *et al.*, 1974).

Tracer injections into either the rostral field or the AI of owl monkeys result in labelled terminals in both fields on the opposite side and in a rostromedial area (Fitzpatrick and Imig, 1980). Tracer injections in marmosets made at frequency-defined sites in the AI also lead to terminal labelling in several patches in and adjacent to the contralateral AI.

The corpus callosum is absent in marsupials and monotremes, but interhemispheric connectivity between the auditory cortex of each side occurs through the anterior commissure and the *fasciculus aberrans*, through the thalamus (Heath and Jones, 1971). Injections of retrograde tracer at frequency-specified sites in the AI of possums (Aitkin and Gates, 1983) and marsupial native cats (Kudo *et al.*, 1989) label cells in the opposite AI.

Thus, in all non-human species so studied, the AI defined by physiological criteria receives input from at least the opposite AI. It is not clear how dense these projections are, compared with ascending thalamocortical connections, so that the functional efficacy of interhemispheric connections remains to be established. However, it has long been known that impulses from both callosal and thalamocortical fibres in cats can converge upon the same neurones, and that the callosal input can either facilitate or depress the cortical responses evoked by clicks (Bremer, 1958; Chang, 1953).

One major difference between humans and other mammals is the great development of oral language for communication. A major consequence of callosal section in humans is the inability to access language (usually a left hemisphere function) to describe right hemispheric perceptions or events. However, there is little evidence to suggest that aspects of non-verbal hearing *per se* are altered in humans by callosal section. Gazzaniga (1970) reports that split-brain patients perform normally in a localization task when sounds emanate from one of a semicircle of small speakers around a subject. It is only when the sounds are words, or part words, that interference with comprehension or memory occurs in split-brain patients. This could suggest that the corpus callosum in humans may have different connections in the temporal lobes to those of lower mammals.

INTRAHEMISPHERIC CONNECTIONS

We have shown that the different auditory cortical fields receive input from different combinations of neurones located in the four principal divisions of the auditory thalamus. This differential connectivity maintains a template for parallel information processing originally manifested in the brain stem. Interhemispheric projections must also be integrated into this cortical framework, and we will now examine how the different cortical areas in one hemisphere connect to each other.

Early studies in cats by Downman *et al.* (1960) showed that electrical stimulation of foci in the AI produced well-localized short latency responses in secondary areas indicative of direct intracortical connections between these areas. Later studies of degeneration consequent upon small lesions to parts of the auditory cortex (e.g. Diamond *et al.*, 1968b; Kawamura, 1973) must be viewed with some caution because the lesions were localized on the basis of sulcal boundaries, which have been shown to vary in cats in relation to mapped auditory fields (e.g. Merzenich *et al.*, 1975). Such lesions will interfere with fibres of passage as well as those of local origin. However, there is concordance between the results of these studies, which show specific interconnections between areas AI, AII and the posterior fields.

Greater precision is achieved with tracer injections made into frequency-mapped areas; these studies have been carried out in cats (Imig and Brugge, 1978; Imig and Reale, 1980, 1981). Topographic, divergent projections occur from the AI to the three tonotopically organized fields in the same hemisphere. In the reverse direction, a single injection into a tonotopically organized field results in multiple terminations in the remaining three fields along the appropriate isofrequency contour. In the owl monkey, similar patchy connections can be observed between the fields AI and R (rostral), and with other less well defined fields

(Fitzpatrick and Imig, 1980, 1982). There are also projections from the AI to non-primary fields in the marmoset (Aitkin *et al.*, 1988a) and rhesus monkey (Seltzer and Pandya, 1978).

EFFERENT PROJECTIONS

What are the routes along which information processing, carried out in the auditory cortical fields, can influence perception and behaviour? If one considers perception as a cortically related phenomenon, one would expect that projections to non-specific or association cortical areas should be present. Such connections do exist in cats (Diamond *et al.*, 1968b; Kawamura, 1973) to pericruciate, anterior lateral and medial suprasylvian association areas, where approximately 70% of cells are responsive to acoustic stimulation. The majority of these cells are also driven by visual and/or somatosensory stimuli (Irvine and Huebner, 1979). These association pathways have been traced further to a subcortical structure, the claustrum (Irvine and Brugge, 1980). In monkeys, in contrast to cats, projections also reach parts of the frontal cortex (Jones and Powell, 1970; Pandya *et al.*, 1969).

The descending connections to the medial geniculate body have already been described in relation to the concept of thalamocortical loop circuits. Longer loops also occur involving the inferior colliculus. On the basis of tracer injections placed into the AI, anterior field and AII at frequency-defined sites in cats, Andersen *et al.* (1980b) have shown that each region projects fibres bilaterally, but particularly ipsilaterally to the dorsal cortex of the inferior colliculi. These projections appear to be in topographic register. In owl monkeys projections from the auditory cortex (putatively the AI) extend well into the central nucleus (Fitzpatrick and Imig, 1978). Similar pathways exist in marmosets (Aitkin *et al.*, 1988a). There may thus be species differences between primates and cats in this projection.

The auditory cortex also sends fibres to premotor areas, although the precise site of origin of these connections cannot be established unequivocally because of the lack of physiological mapping verification. In cats, lesions of an area corresponding to the AII on cytoarchitectural grounds result in terminal degeneration in the superficial layers of the superior colliculus bilaterally, while projections from the posterior cortical fields terminate in deeper layers of the superior colliculus (Paula-Barbosa and Sousa-Pinto, 1973). The AI does not appear to be involved in these pathways. In agreement, lesions of the supratemporal plane of the squirrel monkey, containing the AI, provide degenerating terminals in the MG and IC but not the superior colliculus (Forbes and Moskowitz, 1974).

Area AII, in cats, is also the principal source of the modest projection to the lateral pontine nucleus, a region supplying a major input to the auditory area of the cerebellum (Brodal, 1972). These terminals are associated with those originating from the external nucleus of the inferior colliculus and superior colliculus (Kawamura, 1975). Through these pathways auditory input may modulate motor coordination provided by cerebellar circuitry. Finally, a more direct access for auditory cortical input to motor systems is via pathways to the basal ganglia (Webster, 1961, 1965). On the basis of mapping and tracing studies in cats, Reale and Imig (1983) have shown that the dorsal parts of the putamen and caudate nucleus relate to anterior and dorsal auditory cortical fields, while the ventral putamen and lateral amygdala receive connections from posterior and ventral fields. In any one cortical field, low-CF neurones project medially within the basal ganglia and high-CF cells project laterally. Corticostriate connections may show considerable species variation. For example, in marsupials, much of the auditory inflow to the basal ganglia may originate, not from the auditory cortex, but from the MG (Kudo *et al.*, 1986, 1989).

There are thus differences in the patterns of ascending, cortico-cortical and descending connections of the various auditory cortical fields, although it must be remembered that most studies have been carried out on cats, and only a handful of these studies include unequivocal physiological controls for the cortical area under examination.

There is clearly a connectional substrate for subdivision of auditory cortical areas on a functional basis. Each area receives a different pattern of input from parallel processing channels, the origins of which can be traced back to the auditory brain stem (Figure 5.1). Cortico-cortical connections provide the possibility of a hierarchical processing system with origins within a given cortical area (e.g. Reale *et al.*, 1983). The subject of the intrinsic organization of divisions of the auditory cortex will now be pursued.

6 The internal organization of the auditory cortex

In this chapter I examine the morphology of neurones constituting the auditory cortex, the organization of these cells into layers and columns, and the relationships between the various afferent and efferent projections and the cortical cells receiving and giving rise to these projections. A brief description will also be given of the types of synaptic input observed in the auditory cortex and of the structural and pharmacological correlates of these synaptic actions.

THE LAYERS OF THE AUDITORY CORTEX AND THEIR NEURONAL COMPOSITION

The early Nissl anatomists had shown that the temporal cortex could be divided into layers containing either granule cells (characterized by dense Nissl granules in small cell bodies) or pyramidal cells (larger cell bodies with a triangular shape). One such analysis – that of Campbell (1905) for the human brain – is shown in Figure 6.1a.

Campbell's analysis of the cytoarchitecture of the temporal cortex, viewed with knowledge of the effects of lesions of these regions occurring naturally with humans and by experiment in animals, led him to propose the existence of an audito-sensory area and an audito-psychic area – the latter having more complex functions in hearing (see Figure 1.1). While the status of this dichotomy is now in doubt, his Nissl analysis of the putative primary area in humans is very similar to a modern account for cats, given by Winer (1984a) (Figure 6.1b) although the allocation of cell types into cortical layers differs between the studies.

Rose (1949) made an analysis of the cytoarchitecture of the auditory cortex of cats in material that was also used to demonstrate the connections of the various auditory fields with the MG (Rose and Woolsey, 1949). In this important study, Rose discussed the philosophies underlying the approaches used in architectonic researches, and argued that the definition of a field tacitly assumed by early anatomists – a sector possessing certain consistent characteristics – ignored the common variation of characteristics within a field. These variations are usually

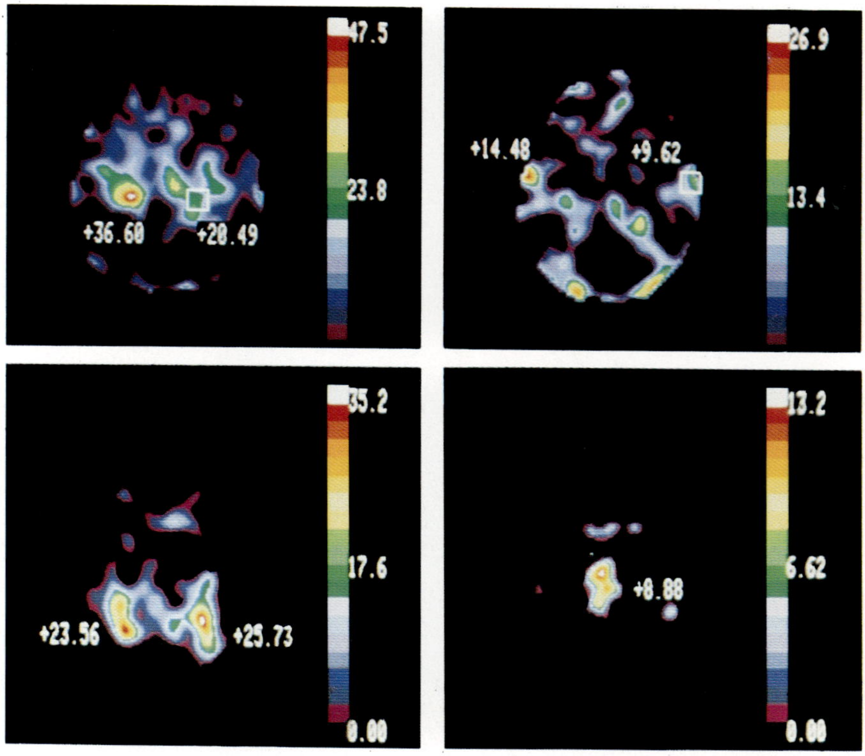

Figure 2.2 Four 'slices' of the human brain, from dorsal and rostral (top left), to more caudal (bottom right), obtained from positron emission tomography of a human subject listening to synthetic syllables presented binaurally at 20 per second and 50 dB sound pressure level. The colour code bars express regional cerebral blood flows obtained by comparison with a control; note that the maximum blood flow varies from slice to slice. Data collected on the PETT VI at Mallinckrodt Institute of Radiology, Washington University at St Louis using intravenous ^{15}O and 40 s scan time. Modified from Lauter *et al.* (1988).

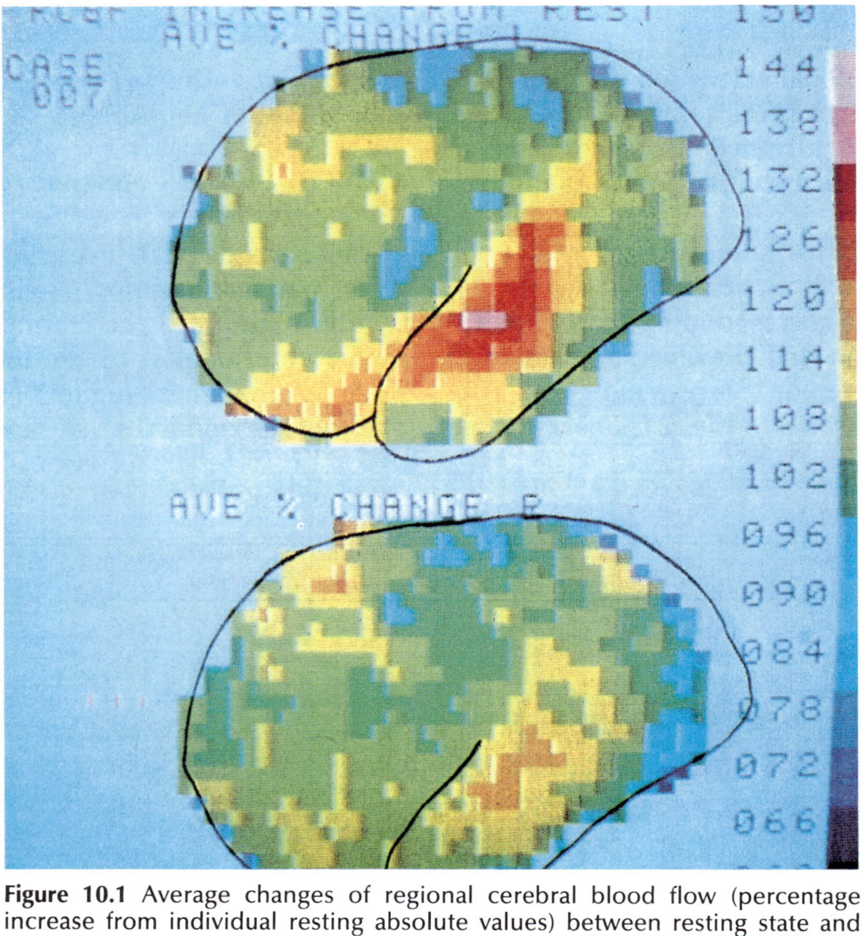

Figure 10.1 Average changes of regional cerebral blood flow (percentage increase from individual resting absolute values) between resting state and verbal stimulation, the latter consisting of 20 simple words presented over about 40 seconds at a distance of 1m from one ear of the subject (the other was plugged). Upper panel: left side; lower panel: right side. From Nishizawa *et al.* (1982).

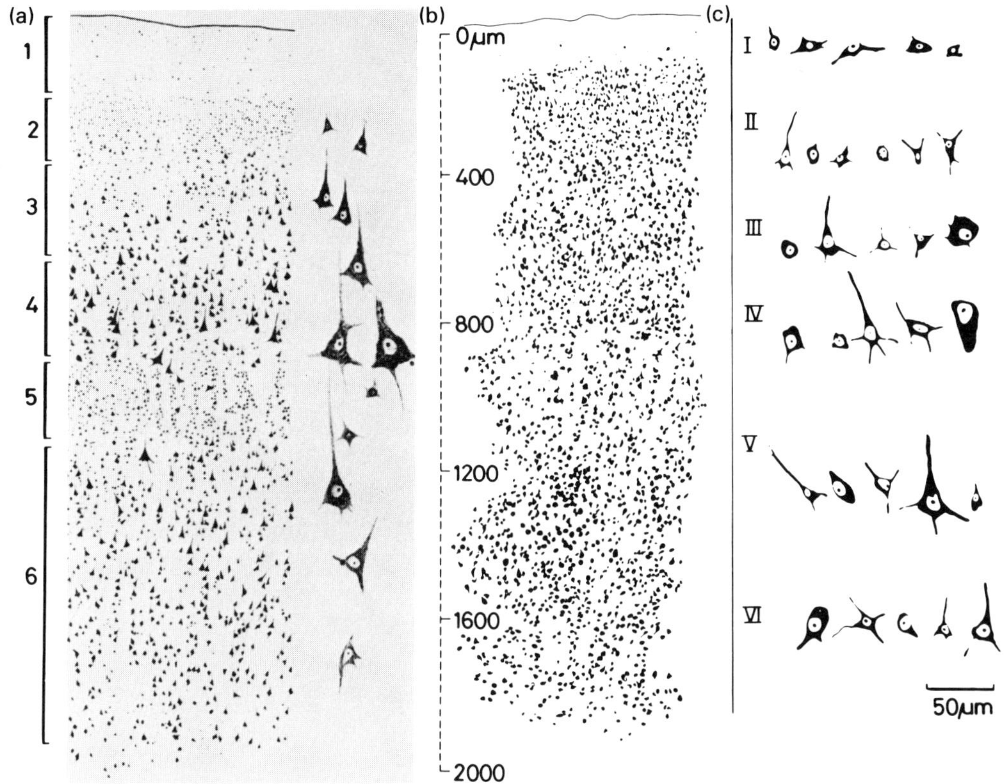

Figure 6.1 Lamination of the primary auditory cortex revealed by Nissl stains. (a) Human 'audito-sensory' area, from Campbell (1905). Campbell divides the cortex into six layers, with two granular layers (2 and 5). Layers 3 and 4 are distinguished by the size of the pyramidal cells (right side of (a)). (b) The AI of a cat, from Winer (1984a). Variation in packing density is obvious, as in (a). Unlike Campbell, but in keeping with the later observations on the human koniocortical area TC of Von Economo (1929), layer III is equivalent to layers 3 and 4 of Campbell, layer IV is densely packed with small cells and is equivalent to layer 5 of Campbell, while the latter's layer 6 becomes subdivided into V and VI of Von Economo and his successors, including Winer (1984a).

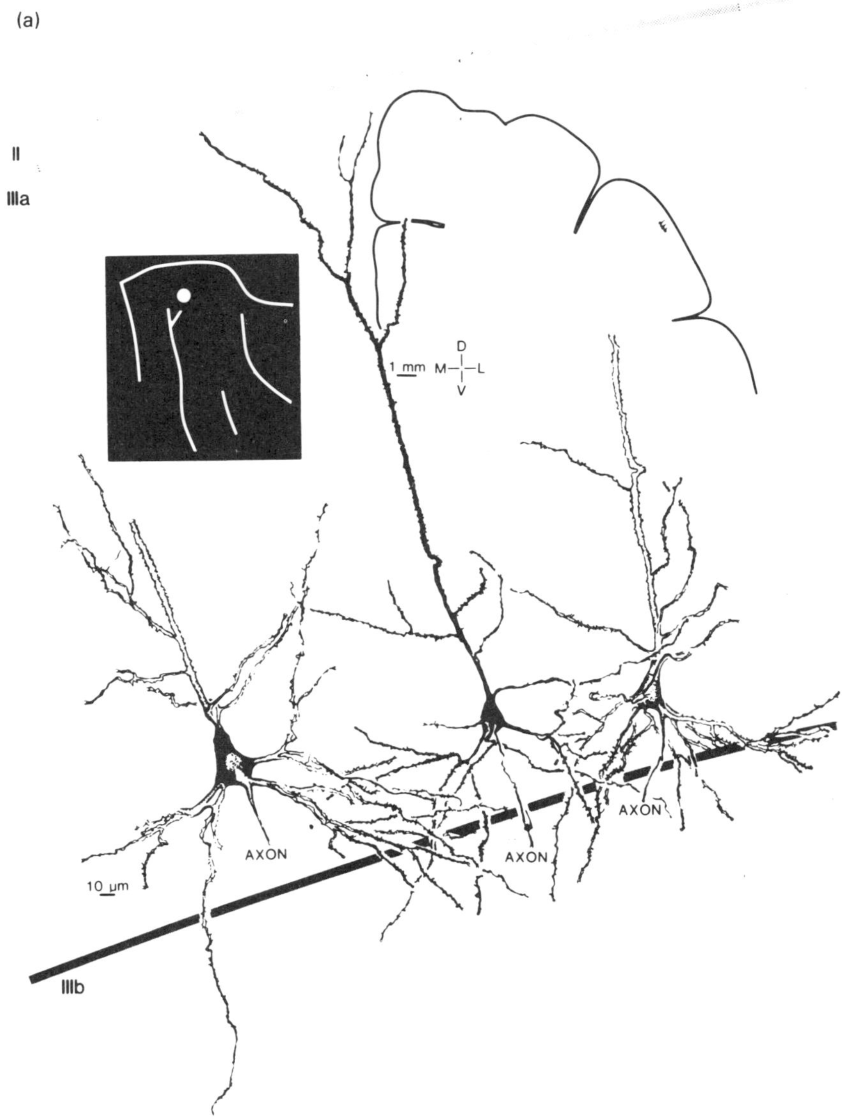
(a)
II
IIIa
1 mm
D
M
L
V
AXON
AXON
AXON
10 μm
IIIb

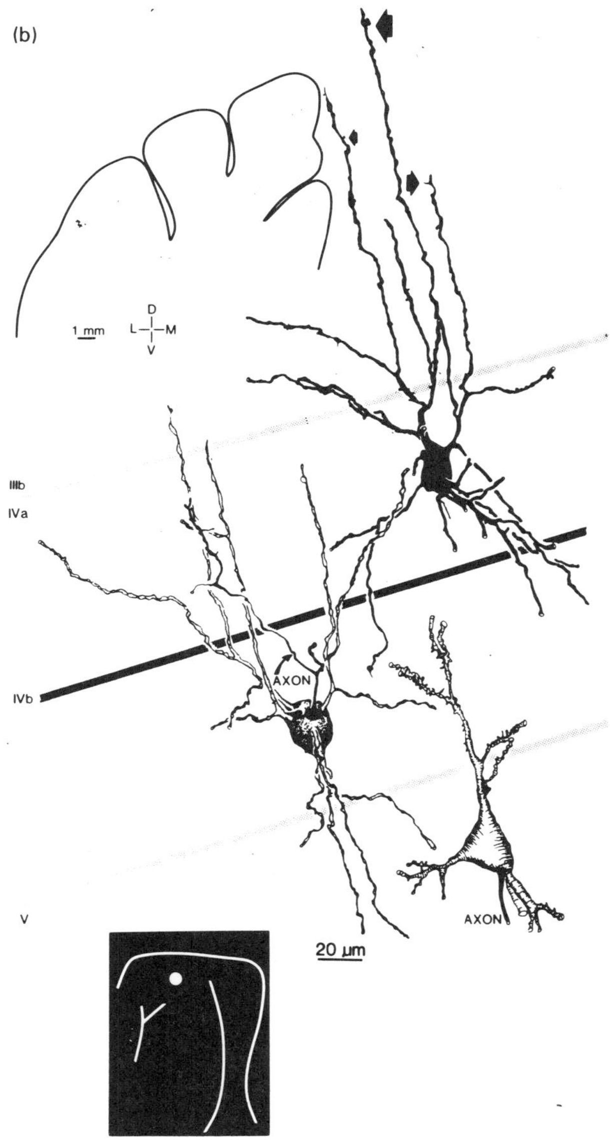

Figure 6.2 (a) Pyramidal neurones (layer III), from Winer (1984b). (b) tufted neurones (layer IV), from Winer (1984a). Insets: line drawings – outline of frontal sections of cat's cortex showing location of stained segment in that plane; dark inset – surface view of lateral part of cat's cortex with major suprasylvian and ectosylvian sulci in white; dot is location of stained segment in that plane.

systematic and are referred to as gradients of a field.

Rose considered the auditory region of cats 'as consisting morphologically of a central field (AI) possessing a characteristic structure of its own, and a peripheral belt which represents structurally a transition zone' (Rose, 1949, p.422). The characteristics of neurones in the AI, but not in other fields, have been studied extensively by Winer and co-workers, so it is of interest to examine the gradients that Rose observed in the AI in Nissl material. First, layers II, III and IV are readily distinguishable in the dorsal part of the field and fuse together in other parts of the field, giving a darker blurred lamination pattern. Layer V is essentially a cell-sparse zone with occasional large cells, some of them the largest cells in the AI. Layer VI contains medium-sized cells fairly densely packed, and is uniform throughout the AI. The cells of the AI are arranged in well-marked columns, particularly in layers III and IV. A recent cytoarchitectural study by Winer (1984a) reaffirms these observations of Rose, although Winer states that there is always a clear distinction between layers II and III.

Nissl stains reveal the shape of the cell body, its nucleus and nucleolus and portions of proximal dendrites. A comparison of the types of neurones revealed by this staining method from the early work of Campbell (1905) and the recent observations of Winer (1984a) show essentially the same picture (Figure 6.1), although, as indicated above, the allocation of neurones to layers has been substantially revised since Campbell's time, his layer 5 being equivalent to Winer's layer IV. The use of the Golgi technique has revealed the variety of cell types subsumed under the general terms pyramidal and stellate. Let us examine the detailed Golgi analysis made by Winer of the auditory cortex of cats.

Pyramidal cells are found in all layers of the AI – those in layer II are very small, and some in layers III and V are very large – but their cell bodies usually lie outside of layer IV. Some examples of these are shown in Figure 6.2a, located in layer III (Winer, 1984b). In addition to their pyramidal-shaped cell bodies, they are characterized by a dominant long apical dendrite arising from the apex of the pyramid and a large number of horizontally spreading shorter dendrites originating from the base of the pyramid. All dendrites feature dendritic appendages, known as spines, which appear short and stubby and receive synapses from axonal terminals of varied origins. Basal dendrites of adjacent neurones overlap extensively, while apical dendrites may reach as far as layer I. The dominant radial ascending orientation of the apical dendrites of cortical neurones gives a columnar appearance to low-power views of the auditory cortex stained by the Golgi method (Figure 6.3).

Stellate cells are generally smaller than pyramidal cells and have relatively spherical cell bodies with dark Nissl substances. They are

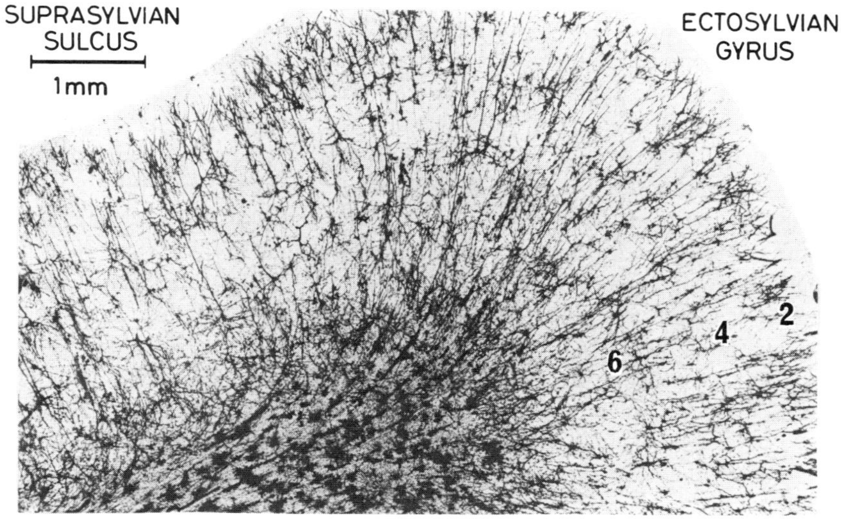

Figure 6.3 Low-power view of the dorsal part of the ectosylvian gyrus containing the AI, in a cat. Golgi–Cox stain. Note how the dendrites of neurones radiate out from the white matter (stained densely and in clumps in the lower left of the picture) to give the cortex its characteristic columnar appearance. Layers 2, 3 and 6 shown for orientation.

probably co-extensive with the granule cells of early neuroanatomists. A similar type of neurone, common in layer IV of the AI, is the tufted neurone of Winer (1984a) (Figure 6.2b). Tufted neurones have few spines, and have localized dendritic fields that tend to emerge from only a few points on the soma surface. Multipolar and spiny stellate cells also have restricted dendritic fields but these are more spherical than those of the tufted neurones.

It is not an easy task to summarize the neuronal populations in the different layers of the AI. Within each major class (pyramidal and stellate) there are many variants and it is not clear if these variants belong to a continuum or represent functionally different types of neurone. There are also neurones that do not appear to belong to either class, such as the double-bouquet cell, the dendrites of which arise from two poles of a soma flattened in the radial plane, and spread for considerable distances as two tightly collected bundles running towards and away from the pia.

The structure of the human auditory cortex (areas 22, 41 and 42) has been described by Seldon (1981a,b), who finds a similar variety of dendritic tree conformations to those seen in cats, including pyramidal, smooth stellate and double-tufted neurones. Neurones are arranged in vertical columns that extend from layer V to layer II, the column width being about one cell wide except in layer IV where two or three cells group together in a column. Columns in the left hemisphere are wider, and separated by more neuropil, than those in the right hemisphere. However, at least for areas 41 and 42, the overlap of dendrites in the tangential direction is much the same on each side. This suggests that a larger amount of neuropil, and probably therefore synaptic machinery, is available for processing in the left than in the right auditory cortex of humans. It has also been shown that, in humans, there are gross asymmetries between the left and right temporal lobes of the part of the superior temporal plane posterior to Heschl's gyrus, involving portions of areas 42 and 22, and corresponding to Wernicke's speech area. This region is larger on the left hemisphere than on the right in 65% of humans, and longer on the left side than on the right side (Geschwind and Levitsky, 1968).

Let us summarize the laminar organization of the AI from the point of view of Winer's (1984a) observations on cats:

Layer I: cell-poor zone, axons superficially and terminal dendrites deeper.
Layer II: small cells, many pyramidal.
Layer III: pyramidal, tufted and multipolar cells. Large pyramidal cells deeper in this layer.
Layer IV: small cells in superficial part, larger cells deep. Few or no pyramidal cells, stellate cells predominate.
Layer V: two sublayers. A superficial cell-sparse layer and a deeper cell-rich layer with some large pyramidal cells.
Layer VI: complex, wide, many pyramidal cells, some of which are horizontal or even inverted.

LAMINAR ORGANIZATION OF PATHWAYS RELATED TO THE AI

We have described the various pathways afferent and efferent to the AI, and the laminar structure and neuronal composition of this region. How do the individual layers differ in their participation in these pathways? Retrograde and anterograde tracing techniques and electrical stimulation experiments have proven crucial to the unravelling of the circuitry involving the primary auditory cortex in cats.

The bulk of thalamic input to the AI terminates in layers III–IV (Aitkin, 1976; Wong and Kelly, 1979). In squirrel monkeys, axon terminals occur on the soma and proximal dendrites. Synaptic appositions, whether on somas or dendrites, are usually of the asymmetric type and terminals have round vesicles, suggesting a predominantly excitatory mode of action (Smith and Moskowitz, 1979). Although additional terminals are observed in layers I and VI, layer IV may be distinguished as the principal thalamo-recipient region and thus the primary origin of intracortical projections of the AI. Injections of retrograde tracers into AII, into the contralateral AI and into the MG or inferior colliculus, label very few cells in layer IV (Code and Winer, 1986; Kelly and Wong, 1981; Winguth and Winer, 1986). A combined electrophysiological and labelling study by Mitani *et al.* (1985) has shown that layer IV cells, all non-pyramidal, are excited at short latency by MG stimulation, confirming their direct thalamic input, but are not antidromically activated from AII, Ep or contralateral AI, confirming their lack of projection to these areas. These data would suggest that layer IV cells project mainly to sites within the AI.

Only a small number of layer II cells are labelled retrogradely when injections of tracer are made in the ipsilateral or contralateral auditory cortical areas (Code and Winer, 1985; Winguth and Winer, 1986). The axons of stellate neurones may arborize locally and those of pyramidal cells can reach layers V and VI. The electrophysiology of these neurones suggests that they are parts of processing loops within the AI (Mitani *et al.*, 1985).

Layer III neurones, mostly pyramidal, are responsible for 75% of the axons travelling from the AI in the corpus callosum (Code and Winer, 1985; Imig and Brugge, 1978), and for a large proportion of ipsilateral cortico-cortical fibres (Winguth and Winer, 1986). Excitatory input to certain layer III neurones originates from the ipsilateral MG, contralateral AI and ipsilateral AII and Ep (Mitani *et al.*, 1985). Layer III is clearly an important source of extracortical projections that carry information derived from several intra- and extracortical loci.

Large pyramidal cells in layer V, like those in layer III, are responsible for an important part of the cortical outflow of the AI. In addition to their projections to ipsilateral and contralateral auditory cortical areas, these cells are the source of many of the axons descending to the MG and IC (Kelly and Wong, 1981). In both cats (Mitani and Shimokouchi, 1985) and rats (Games and Winer, 1988) the pyramidal cells giving rise to cortico-collicular projections are segregated from those projecting to the MG or cortical areas; however, in rats the cortico-collicular projections arise from cells located deep in layer V, while those in cats are mostly in the upper part of this layer. This could either be a species difference or relate to different definitions of cortical layer boundaries. Electrical stimulation of the MG antidromically activates many neurones in layer V and excitatory inputs originate from a number of ipsilateral and contralateral auditory cortical regions, as well as from the ipsilateral MG (Mitani *et al.*, 1985).

Layer VI cells also provide axons to the corpus callosum and ipsilateral auditory cortical regions, but they are not major contributors to these pathways. The pyramidal cells are, however, important sources of descending axons to both the MG and the IC (Kelly and Wong, 1981; Mitani *et al.*, 1985). Non-pyramidal neurones of layer VI may be excited at short latency by stimulation of the MG, but do not appear to project out of the AI (Mitani *et al.*, 1985).

Matsubara and Phillips (1988) have examined some of the intrinsic connections of the AI of cats by first mapping the CFs and binaural properties of units in a patch of the AI and then making microinjections of tracers in some portion of this mapped area. They find that up to five patches of labelled terminals and neurones can be observed in the AI around the injection site. The labelled material is found in the middle layers, suggesting that intrinsic connections may be made within a lamina. The patches are found in regions with similar or higher CFs, never lower. There is no correlation with binaural properties, so that injections made in binaural excitatory areas can lead to patches of labelling in either binaural excitatory or inhibitory areas. Similar patches are found with injections into the visual cortex, and the authors speculate 'that a universal, anatomical plan of local connections may exist within the neocortical mantle'.

SYNAPTIC ACTIONS ON AI NEURONES OF DIFFERENT INPUT PATHWAYS

What synaptic actions do the various inputs have upon target neurones in the AI? This question has been addressed by Mitani and co-workers (Mitani and Shimokouchi, 1985; Mitani *et al.*, 1985), who stimulated the MG, IC, contralateral AI and ipsilateral AII and Ep electrically and recorded intracellular potentials in the AI.

Excitatory postsynaptic potentials (EPSPs) were recorded in 91% of neurones in the AI with MG and IC stimulation. The latencies of these EPSPs indicated the mono- or polysynaptic route taken by evoked impulses, with the shortest latencies found at a depth corresponding to layers III–IV and the longest in layer I. In other neurones, found throughout the AI, both short- and long-latency EPSPs were evoked. The longer latencies of IC versus MG monosynaptic EPSPs were interpreted as being a result of relay in the ventral division of the MG. Inhibitory synaptic actions were also revealed by MG and IC stimulation, with inhibitory postsynaptic potentials (IPSPs) succeeding EPSPs with a delay of about 1 ms in about 34% of impaled neurones.

For many neurones, EPSPs were evoked from more than one stimulation site. For 38% of neurones short-latency EPSPs were evoked by MG,

AII and Ep stimulation, indicative of monosynaptic activation from each site. In general, combinations of short- or long-latency input were observed, with different neurones receiving different combinations of input. Excitatory postsynaptic potentials were much more common than were IPSPs when the AII, Ep and contralateral AI were stimulated.

This first step in pathway tracing using electrophysiological methods has revealed something of the complexity of the synaptic actions on neurones in the AI. The question of the neurotransmitters responsible for these actions is only beginning to be addressed in detail, and the applications of immunocytochemical techniques and of receptor binding methods to the auditory cortex are still in their infancy. These techniques are necessary in order to determine which putative transmitters occur in potentially useful levels at synapses, or within neurones, of the AI. This information needs then to be corroborated by neuropharmacological investigations of the synaptic actions of these substances and their antagonists.

One likely candidate for inhibitory transmission in all sensory cortices is gamma-aminobutyric acid (GABA). Winer (1986) has shown in cats that GABA has a widespread distribution in the supragranular layers of the AI. Neurones which are positive for GABA tend to be small and of varied dendritic morphology, but none of the labelled somas have a pyramidal shape. This could suggest that GABA has an entirely interneuronal function were it not for the observation (Winer, 1986) that GABA is taken up by some cells projecting to ipsilateral and contralateral auditory fields. A study by Foote *et al.* (1975), employing iontophoretic application of putative neurotransmitters, reported that the spontaneous and sound-evoked activity of cells in the superior temporal gyrus of the squirrel monkey were inhibited by GABA. All cells studied were inhibited, and the inhibition had a rapid time course compared with that of noradrenaline, which was also inhibitory. This study, which additionally showed a common excitant action of acetylcholine, needs to be replicated with more histological controls, in cats. However, it is clear that there are cells cytochemically positive to GABA in the auditory cortex and that GABA has an inhibitory action in this region.

The importance of GABAergic inhibition in auditory forebrain processing has been demonstrated clearly for a bird, the domestic fowl (Müller and Scheich, 1988). Iontophoresis of the GABA antagonist, bicuculline, expanded the range of frequencies that excited auditory forebrain units and suppressed the inhibitory influence normally provided by pure tones with frequencies lying outside the excitatory response area. Müller and Scheich propose that GABA is an inhibitory transmitter with at least two roles in the auditory forebrain – one to provide non-monotonic behaviour as a function of intensity and the other to limit the width of the tuning curve.

McKenna *et al.* (1988) have examined the actions of micropressure injections of acetylcholine and methacholine on units in the AI of paralysed cats. They found that the evoked activity of 80–90% of units was influenced by these agents, with similar numbers being facilitated or suppressed. Many of these effects were blocked by atropine, which often had depressant effects of its own. All actions had a long latency of onset and outlasted the applications by many minutes. In general this study is not supportive of a neurotransmitter role for acetylcholine in the AI, but the use of the nicotinic anticholinergic agent, gallamine triethiodide, to paralyse the cats may have interfered with the responsiveness of units to applied cholinergic agents.

COLUMNAR ORGANIZATION OF THE AI

As a conclusion to this chapter, let us look at the structural evidence for the existence of 'columns' of radially interconnected neurones in the AI. This will also be a focus in the next chapter when physiological response properties of cells in the auditory cortex are examined.

It is obvious in Nissl-stained sections of the auditory cortex that cells, particularly in layers III and IV, are arranged in strings perpendicular to the pial surface and these are separated by columns of fibres. How discrete are these columns? What is the extent of territory sampled by terminal arborizations of thalamocortical axons, and how far do the dendrites of cortical neurones extend?

Demonstration of terminal branching patterns of afferents to the AI ideally requires the use of intra-axonal injections of horseradish peroxidase, as has been used in the visual cortex (e.g. Gilbert and Wiesel, 1981). However, silver and Golgi stains show that radial bundles of fibres separate the columns of cells (Winer, 1984a,b), giving a column width of approximately 50–75 μm. The apical dendritic fields of pyramidal cells and the total fields of the smaller stellate cells fall within this column width (or width-and-a-half, since the cell bodies are flanked with columns of neuropil on each side). However, the dendritic fields of the larger stellate cells and the basal dendrites of pyramidal cells extend well outside the column width, and there is a considerable horizontally disposed fibre plexus in the AI, particularly in its middle and deeper layers (Glaser *et al.*, 1979; Winer, 1984a). Winer (1984a) depicts a layer IV 'microcolumn' as a rectangular prism of dimension 50 μm by 200 μm, which includes most of the major cell types he describes and substantial parts of terminal fields of large, presumed-extrinsic axons. He further relates these cortical microcolumns to the fibrodendritic laminae observed in the ventral division of the MG (Morest, 1965).

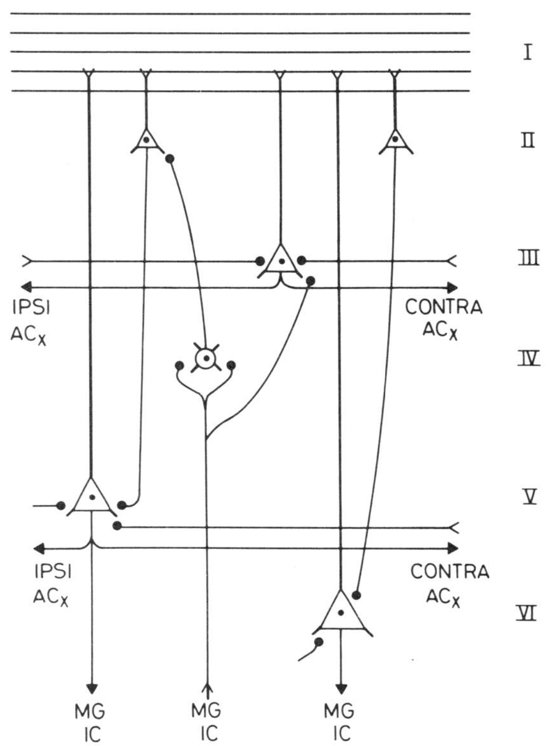

Figure 6.4 Idealized circuitry of an auditory cortical microcolumn. Synaptic terminals are shown as black knobs and are found on small pyramids of layer II, medium pyramids of layer III, large pyramids of layers V–VI, and stellate (tufted) neurones of layer IV. ACx, auditory cortex; MG, medial geniculate body; IC, inferior colliculus.

The circuitry of the microcolumn is idealized in Figure 6.4. Here, no attempt has been made to include the stellate cells (except for the thalamo-recipient cells of layer IV); these appear to function mainly as intra-AI interneurones. The connections between layers II and V–VI are hypothetical. Note that most of the output of the microcolumn will be carried by pyramidal cell axons that run *within* the column; connections from, and to, other cortical areas are shown as laterally directed only for convenience of illustration.

In his Golgi studies of the auditory cortex of humans, Seldon (1981b) observes a banded pattern of fibres in both the sagittal and horizontal planes. In tangential sections bands of fibres about 200–300 μm wide weave an irregular path through each auditory area (areas 22, 41 and 42) and are separated from neighbouring macrocolumns by similar dimensions. Seldon speculates that these bands may be callosal in origin and related to neurolinguistic phenomena. It is not clear how many microcolumns constitute a macrocolumn; it is in fact possible that the two concepts are one.

7 Physiological response properties of neurones in the auditory cortex

In this chapter we will examine how neurones in the auditory cortex respond to the intensive, spectral and temporal characteristics of sounds. In all species, one important role of the auditory pathway is to analyse these characteristics in order to identify sounds of particular importance to that species. The ultimate aim of such analysis is to identify 'what' the sound is and what information it might convey about a con-specific or about potential predators or prey. The methods used are essentially those of stimulation with pure and complex sounds and recording the responses of single neurones to those sounds with microelectrodes (see Aitkin *et al.*, 1984b).

It has been common for parametric studies of the physiological responses of auditory neurones to employ sound stimuli delivered from transducers sealed into one or both external auditory meatuses. Such an approach allows for the accurate calibration of sound pressure levels at the eardrum, but important influences of the external ear on the sound field, that would occur with sounds in the free field, are lost. This problem is more acute in studies of binaural localization, where both ears contribute input to neural coding mechanisms; questions of 'where' the sound is located will be addressed in the next chapter.

The early study by Erulkar *et al.* (1956), which in fact employed free-field acoustic stimuli, surveyed the types of auditory cortical responses that may be recorded with microelectrodes. They established the criteria for 'unitary potentials'; the extracellular manifestations of action potentials of single neurones. It is now conventional to refer to the source of these all or nothing spikes as 'units'. They examined the distribution of units in the AI and showed that responsive units could be found at all depths, although they were most common in layers III–IV. They examined the question of anaesthetic depression of responsiveness and varied the type of anaesthetic employed. They concluded that while deep anaesthesia clearly depressed unit responsiveness, unresponsive and responsive units could be encountered in the immediate vicinity of each other.

Questions of anaesthetic effects have remained an important issue since this germinal study by Erulkar *et al.* It is natural to question cortical activity correlates of perception in conditions where perception does not occur in humans. Gerstein and Kiang (1964) used unanaesthetized, unrestrained cats and free-field stimuli, but did not comment on the proportion of unresponsive units. Using a similar technique, Evans and Whitfield (1964) found that about 20% of units were unresponsive to acoustic stimuli. Goldstein *et al.* (1968, 1970), who used a muscle-relaxed, locally anaesthetized preparation, found that most units responded to some acoustic stimulus. These techniques for recording from unanaesthetized cats have not been used in recent years because of ethical considerations, and, in the case of the unrestrained cat, the difficulty in controlling the stimulus intensity and state of arousal of the animal. Most parametric work on cortical coding continues to rely on anaesthetized animals which offer stable, albeit abnormal, recording conditions for long periods of time. An underlying assumption of these experiments is that anaesthetic depression is not likely to be selective on frequency or intensity ranges but should simply reduce the number of action potentials a given stimulus will evoke from a unit.

In order to present a picture of the physiological properties of units in the auditory cortex that coheres with the foregoing account of cyto-architecture and connectivity, this review must be selective. I will give closest attention to parametric studies of the coding of simple acoustic stimuli in which electrophysiological or histological criteria are given for the location of unit recording sites. Most of the parametric data have been obtained from the AI of cats, but observations made in other species, or auditory fields, will be presented where relevant.

RESPONSE PATTERNS TO SIMPLE ACOUSTIC STIMULI

In anaesthetized animals most units in the auditory cortex respond to tonal stimuli, clicks and noise with one or two action potentials at the onset of the stimulus (Brugge *et al.*, 1969). This pattern is also very common in unanaesthetized preparations (Evans and Whitfield, 1964; Goldstein *et al.*, 1968). Such onset responses are associated with excitatory postsynaptic potentials (EPSPs) and these are often followed by inhibitory postsynaptic potentials (IPSPs) (de Ribaupierre *et al.*, 1972). This suggests that acoustic input leads to inhibitory activity that is at least partly intracortical in origin. Given the reciprocity of connections between the AI and MG, and the presence of intrinsic inhibitory mechanisms in the MG, the role of inhibition in modulating cortical activity is clearly very important.

Auditory cortical units also fire spontaneously, in the absence of overt acoustic stimulation, and intracellular recordings show the presence of spontaneous EPSPs and IPSPs (de Ribaupierre *et al.*, 1972). Spontaneous firing occurs at very low levels in anaesthetized animals. It has been argued (Webster and Aitkin, 1971) that barbiturate anaesthesia enhances and prolongs inhibitory activity in the thalamocortical auditory system, leading to a generalized reduction in firing of both evoked and spontaneous spikes.

In unanaesthetized animals responses occurring during the presentation of a prolonged stimulus ('through' or 'sustained') are also observed in a significant proportion of units. These may take the form of sustained excitation or sustained inhibition (of spontaneous activity) and may be associated with a preceding onset response. 'Off' responses at the end of a prolonged stimulus are found in about one third of the population (Goldstein *et al.*, 1968). These patterns are very uncommon in anaesthetized preparations.

INFLUENCE OF STIMULUS INTENSITY ON RESPONSE RATE

Most cochlear nerve fibres increase their firing rate as stimulus intensity is increased; firing rate reaches a maximum level that is unaltered by increases in intensity (see Chapter 3). This sigmoidal relationship is called a monotonic function, and is also found with many units in the central auditory pathway, including the auditory cortex.

Analysis of the effects on auditory cortical units of tone or noise intensity changes has been an important aspect of the work of Phillips and co-workers (Phillips, 1987; Phillips and Cynader, 1985; Phillips and Irvine, 1981, 1982, 1983; Phillips and Orman, 1984; Phillips *et al.*, 1985b). Monotonic intensity functions (Figure 7.1a) characterize about 55% of units in the AI and non-monotonic functions are found in the remainder (Figure 7.1b). Approximately 70% of units in the anterior field have monotonic functions (Phillips and Irvine, 1982). Non-monotonic functions reach a peak firing rate and then decline at higher levels, sometimes to zero. Phillips *et al.* (1985b) use a criterion of a reduction to at least 50% of the maximum firing rate to define non-monotonicity. In their sample of units stimulated monaurally under closed-field conditions, the dynamic range (intensity range over which firing rate increases) varied from 10 to 40 dB, values similar to those observed in cochlear nerve fibres (see Chapter 3). It is possible that a class of unit may exist that increases its firing rate beyond this narrow firing range when stimulated under free-field, and thus binaural, conditions.

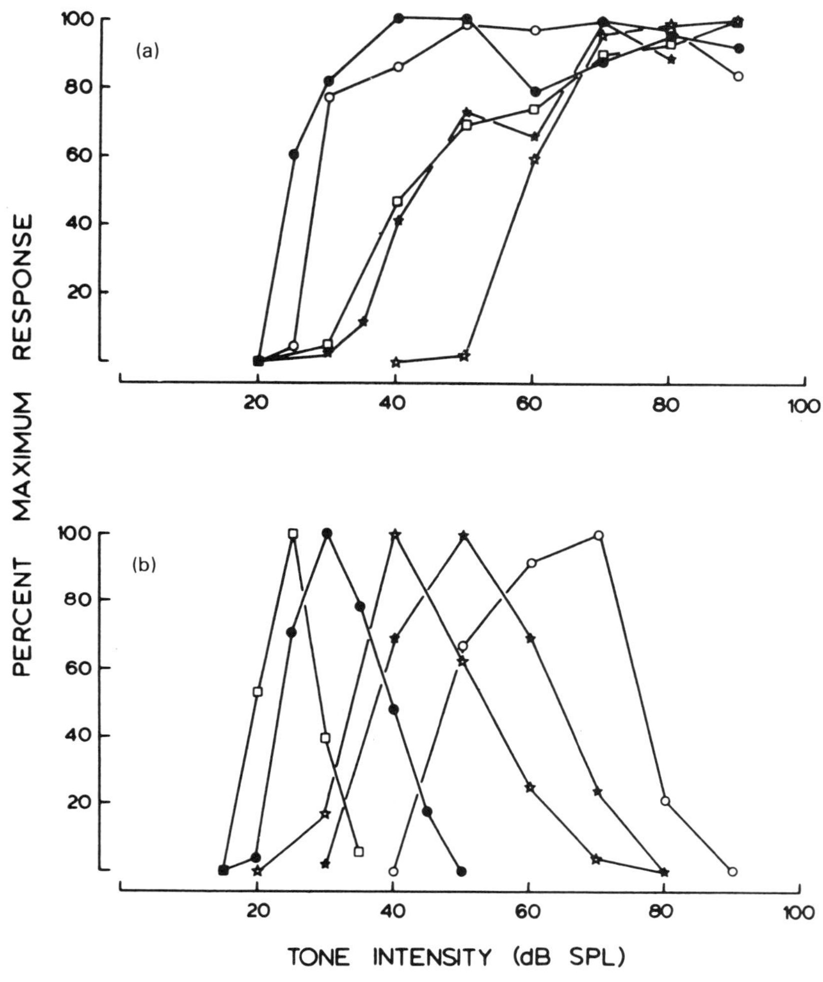

Figure 7.1 Normalized spike count–intensity functions for five representative monotonic (a) and five non-monotonic (b) cells. Data obtained using 50 presentations of CF tones delivered either monaurally to the contralateral ear, or to both ears simultaneously. From Phillips *et al.* (1985b).

Inhibitory mechanisms, probably located peripheral to the AI, must shape the non-monotonic functions. Anaesthetic effects are unlikely to have a dominant influence on such mechanisms, since non-monotonic functions with similar characteristics have been recorded in the AI of unanaesthetized monkeys (Brugge and Merzenich, 1973) and in the MG of awake cats (Aitkin and Prain, 1974). These inhibitory mechanisms operate in the spectral as well as intensity domain. When noise stimuli, containing all frequencies between 20 Hz and 200 kHz, are presented to units with monotonic intensity functions at CF, the noise intensity functions are also monotonic, and the noise spectrum level at the CF needed to evoke activity is substantially lower than that for a pure tone at CF. Phillips and co-workers (Phillips *et al.*, 1985b; Phillips and Cynader, 1985) suggest that this is due to the unit responding to spectral energy in the noise summed across the bandwidth of its frequency tuning curve. In contrast, more than half of the units with non-monotonic intensity functions to tones do not respond to noise stimuli at all. This lack of response occurs in spite of the fact that the spectral energy of the noise at the CF would itself have been adequate to excite the neurone. Phillips *et al.* (1985b) reason that components of the noise at frequencies above and below the CF suppress the excitation that would have been evoked by the CF tone component (see below and Figure 7.2).

Phillips and Orman (1984) studied the responses of units in the posterior field (P) of cats and found a high proportion (86%) of non-monotonic units in this field. For some units the optimal sound level evoking maximum firing varies with frequency in parallel to the threshold tuning curve. This property is also exhibited by units in the inferior colliculus (Aitkin, 1986). For other units the optimal sound level is constant in spite of changes in sound frequency. The latter units may have a role in sound intensity detection since different units require different optimal sound levels. In this case, different sound levels will evoke maximum firing from a varying population of neurones in field P.

Are units with different optimal sound levels organized in a systematic way in fields AI or P? At present, mapping experiments have not been carried out in cats to test this hypothesis, aside from the demonstration (see below) of a segregation of monotonic and non-monotonic units across the AI. However, a detailed analysis of the auditory cortex of mustached bats by Suga and Manabe (1982) has given clear evidence for an orderly relationship between the sound level providing maximum excitation and place within the auditory cortex. Recall (Chapter 4) that a specialized region within the auditory cortex of this species processes frequencies between 61 and 63 kHz. This narrow range corresponds to the second harmonic component of the constant frequency biosonar signal and its Doppler-shifted echoes.

Most of these units have non-monotonic spike count–intensity functions at their CFs, which present sharply defined best sound levels (best amplitudes). As noted in Chapter 4, iso-CF lines in the mustached bat auditory cortex are not linear, but concentric. The distribution of best amplitudes is found to be complex, but orderly in most bats. Best amplitudes shift along iso-CF lines, but are constant as a function of depth normal to the cortical surface, as is CF. Thus, in the mustached bat, there is a disproportionate tonotopic map, an amplitopic map apparently orthogonal to the tonotopic map across the cortical surface, and an in-depth columnar organization where both parameters are constant.

FREQUENCY-RESPONSE CHARACTERISTICS

Pure tone bursts without abrupt onset are perhaps spectrally the simplest stimuli to consider. Tuning curves in cats have been derived for units in the AI (Hind, 1960; Abeles and Goldstein, 1970; Katsuki *et al.*, 1959; Oonishi and Katsuki, 1965; Phillips and Irvine, 1981) and anterior field, A (Phillips and Irvine, 1982). The majority of tuning curves are V-shaped, with little tendency to extend to low frequencies at higher sound pressure levels. Measurements of tuning curve bandwidth indicate that (i) tuning curve sharpness increases on a logarithmic basis from low to high CF but is more uniform on a linear basis and (ii) that on the average, tuning curves are similar in shape and bandwidth to those observed at lower levels of the auditory pathway (Calford *et al.*, 1983). There is little superficially to distinguish between tuning curves from the AI and A in cats (Phillips and Irvine, 1982). Oonishi and Katsuki (1965), in contrast to studies cited above, used binaural free-field stimulation in anaesthetized cats and found a more diverse population of tuning curve shapes, varying from flat, through irregular and multipeaked, to sharp. The defining criteria for shape were not given, but the mean first-spike latencies correlated with sharpness such that flat tuning curves were derived from units with long latencies (18.7 ms) and sharp curves from units with short latencies (10.7 ms).

They examined the relationship between tuning and depth and found (i) that vertical penetrations record units with the same CF and (ii) that flat tuning curves are located superficially, and sharp and multipeaked curves are encountered in the deep cellular layers. Changes in CF are seen when electrode penetrations are oblique to the cortical layers.

This study, performed over 20 years ago, remains of great interest because it is one of the few in which an attempt has been made to analyse the integrative mechanisms of the auditory cortex. Oonishi and Katsuki put forward a model in which MG afferents terminate on sharp (if single tonotopic afferents) or multipeaked units (a convergent input from tonotopically unrelated MG afferents) and these in turn project to units with broad or flat tuning curves. They compared this projection pattern to that put forward by Hubel and Wiesel (1962) to account for different visual receptive fields in area 17 of cats.

One problem with this study is that replication of the observations of tuning curve shape as a function of depth has not been forthcoming (e.g. Abeles and Goldstein, 1972). Most authors find the superficial layers of the auditory cortex unresponsive to tonal stimuli; most activity is obtained from units in the middle cortical layers. Abeles and Goldstein (1970) made a detailed analysis of response characteristics of units in the AI and found that the frequency response ranges of narrowly tuned units were indeed very similar as a function of depth, but that tuning curve shape did not systematically change. It is possible that this discrepancy is related to the use of binaural free-field stimulation by Oonishi and Katsuki (1965) and of closed-field, presumably contralateral, stimulation (although not clearly stated) by Abeles and Goldstein (1970).

Both studies and that of Suga and Manabe (1982) support a concept of columnar organization of the AI with neurones in the same radial cell column possessing the same CF (see also Phillips and Irvine, 1983). This is in accord with both the constancy of receptive field location as a function of depth found for cells in the visual (Hubel and Wiesel, 1963) and somatosensory cortex (Mountcastle, 1957), and the relatively restricted terminal arborizations of ascending afferents in the middle cortical layers of the AI (Winer, 1984a). The possession of a shared input (but not direct connectivity) has also been demonstrated using cross-correlation techniques for neurones lying within 250 μm of each other in the auditory cortex of cats (Dickson and Gerstein, 1974).

Tuning curves define the threshold profile of the connections from the cochlear partition to a unit in the AI, but do not give any indication of the behaviour of the unit at higher intensities. Phillips *et al.* (1985b) have compared the contralateral intensity and frequency response characteristics (response areas) of units in the AI to define two types of response area (Figure 7.2). Units with monotonic intensity functions show very similar tuning characteristics at high and low intensity levels. Neurones with non-monotonic functions, however, have circumscribed response areas such that the frequency–intensity domain of greatest firing is surrounded at higher and lower intensities and frequencies by a zone of decreased excitation. Units with non-monotonic response areas have significantly longer latencies than those with monotonic areas, and the two classes are segregated orthogonal to the tonotopic axis in the middle cortical layers.

Because of the preponderance of non-monotonic units in the 61–63 kHz zone of the auditory cortex of mustached bats, the majority of response areas in this region are extremely circumscribed by inhibitory zones. The net effect is to turn each unit in this area into an extremely sharp filter for frequency *and* intensity (Suga and Manabe, 1982).

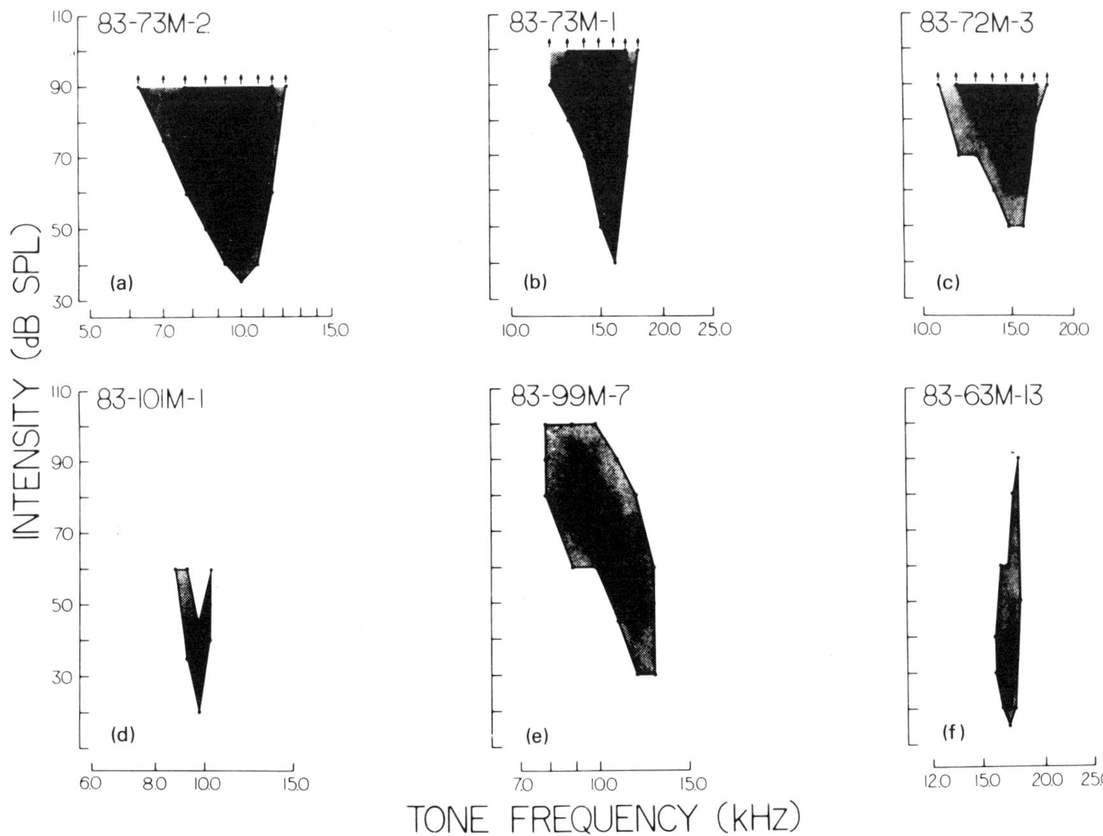

83-73M-2
83-73M-1
83-72M-3
83-101M-1
83-99M-7
83-63M-13
(a)
(b)
(c)
(d)
(e)
(f)
INTENSITY (dB SPL)
TONE FREQUENCY (kHz)

EFFECTS OF TIME-VARYING ACOUSTIC FEATURES

Click stimuli, consisting of transient changes in sound pressure generated by square wave electrical pulses applied to a loudspeaker, were the first types of simple acoustic stimuli to be used by auditory physiologists (e.g. Erulkar *et al.*, 1956; Gerstein and Kiang, 1964). However, the interpretation of the mechanism's underlying responses to clicks is not simple, since clicks contain energy at many different frequencies and this spectral energy will vary depending on the methods of production of the clicks (e.g. Goldstein, 1968, p. 146).

Temporal variations in pure tone stimuli – parametric variations in rise and decay times of tone bursts, and amplitude and frequency modulations of the envelope of the tone burst – offer analysable but complex acoustic stimuli that form a stepping stone to 'real world' communication signals such as speech or species-specific vocalizations. Some units in the auditory cortex appear to respond more strongly to complex stimuli than to pure tones (Evans and Whitfield, 1964; Whitfield and Evans, 1965).

Phillips and Hall (1987) have examined the responses of auditory cortical units in cats to linear modulations in the amplitude of continuous CF tones, and have compared these with responses to steady tone bursts of equivalent intensities. They show that auditory cortical neurones are sensitive to both the sound level of the tone upon which the amplitude modulation (AM) is imposed, and to the rate of the AM. They define an intensity increment response area, relating AM rate and tone level, within which a standard AM increment is able to evoke discharges. Their results are compatible with a model of adaptation in which the presence of a continuous signal (noise or CF tone) leads to an adjustment of a unit's threshold to equal the sound level of the effective component (CF component in the case of noise) of the continuous stimulus.

Phillips (1988) has also examined the influence of stimulus rise time on auditory cortical neurone responses to tone bursts. This could be considered as a special case of AM, from zero to maximum amplitude. However, the onset of a tone burst contains energy not only at the tone frequency but also at other frequencies above and below this frequency. As the rise time decreases towards zero the amplitudes of the 'side'

Figure 7.2 Schematic depictions of the excitatory response areas of three monotonic (a–c) and three non-monotonic (d–f) neurones. Shaded and stippled areas respectively indicate the frequency–intensity domains within which tonal stimuli evoked responses whose strengths were within or less than 50% of maximum response. Arrows in (a–c) indicate that the neurones continued to discharge spikes at intensities exceeding those for which spike count data were collected. From Phillips *et al.* (1985b).

frequencies increase, the onset sounds more like a click, and neurones will be increasingly influenced by frequencies other than the dominant tone frequency.

The general effect of increasing the rise time is to invoke or increase the inhibitory interactions that are manifested in spike count–intensity functions of single cortical units. Although some units with monotonic intensity functions are uninfluenced, other monotonic functions become more bell-shaped, while non-monotonic intensity functions become increasingly so with fast rise time stimuli. Phillips (1988) argues that the increasingly non-monotonic behaviour is explicable in terms of the high- and low-frequency inhibitory side bands that surround the CF region of a response area (see Figure 7.2). Fast rise time stimuli will contain higher amplitudes of these lower and higher frequencies and the neural inhibitory effects they will evoke will interact with events occurring at the CF.

The response of a neurone to frequency modulation (FM) is also partly explicable in terms of the neurone's response area (Phillips *et al.*, 1985a). FM stimulation has been employed in a number of studies of the auditory cortex (Mendelson and Cynader, 1985; Nomoto, 1980; Watanabe, 1972; Whitfield and Evans, 1965). In the earliest of these, by Whitfield and Evans (1965), a sinusoidal modulation of frequency was employed from a free-field speaker to stimulate neurones in the auditory cortex (not necessarily the AI) of unanaesthetized cats. They described some units that responded best to frequency modulations towards the CF, others that responded uniformly to a certain modulation as long as it lay within the response area, and a third group that responded to FM over frequency ranges in which tone burst stimuli were ineffective.

This important early study was germinal for concepts of feature extraction at the cortical level. Whitfield and Evans, in demonstrating (i) responses to FM but not steady tonal signals, and (ii) directional preferences for FM, laid the foundation for later studies of species-specific vocalizations with which FM stimuli have many features in common. However, most of the later studies (to be reviewed below) and that of Whitfield and Evans are difficult to interpret in the absence of detailed analysis of the response areas of units studied with FM, and because of the uncertainty of the auditory fields from which recordings were made.

Such an analysis may be found in the study of Phillips *et al.* (1985a). They used linear FM ramp stimuli of narrow frequency excursion within the excitatory response area and found that the response of a neurone in the AI of cats to such stimuli is related simply to features of the neurone's excitatory response area. Thus, preferences for upwardly directed ramps are observed on the low-frequency slope of the response area and downwardly directed ramps on the high-frequency slope – in all cases, firing increases during the FM stimulus as the CF is approached. In addition, directional preference for a given unit is determined by the steepness of the slopes of the high and low sides of the excitatory response area.

Phillips *et al.* account for these preferences on the basis of the timing of discharges. It is known that the latent period to the first spike is shortest at the CF and progressively longer for tones away from the CF (Aitkin *et al.*, 1970; Brugge *et al.*, 1969). Modulation of the frequency towards the CF will evoke successive impulses of progressively decreasing latency, so that a temporal summation of excitatory events is possible with a resultant increase in firing rate. The presence of inhibitory areas flanking the excitatory area, particularly prominent with non-monotonic units, may lead to asymmetric inhibitory effects so that a null response occurs in one direction of modulation, and a net excitatory response occurs in the other. This accounts for the more varied directional effects of wider FM excursions such as those used by Whitfield and Evans (1965). Such effects would only be revealed with wide modulation ranges, and would not be visible with the narrow excursions employed by Phillips *et al.* (1985a).

AUDITORY CORTICAL UNITS AND 'NATURAL' SOUNDS

An ultimate aim of studying discharge properties of auditory cortical units is to understand how normal communication signals are encoded by these discharges. Species-specific vocalizations contain tonal elements that are both frequency and amplitude modulated, and transients that contain many simultaneously occurring elements. There are a number of points to consider when interpreting studies of unit responses to vocalizations.

First, the various cortical auditory fields may have different roles in vocalization detection – indeed, this process may occur outside the auditory cortex, even peripheral to the cortex. It is therefore important to be able to specify the field(s) from which units under study are drawn. Differences between studies may be due to different fields being sampled.

The depth of anaesthesia, or state of arousal of unanaesthetized animals, will have an important influence on the discharge rate in response to complex and simple stimuli (Pfingst *et al.*, 1977). Responses to a repeated stimulus, no matter how 'natural', may show lability in unanaesthetized animals (Glass and Wollberg, 1979; Manley and Müller-Preuss, 1978). Units responding to a stimulus that causes the performance of a task in an attending animal do so much more vigorously and regularly than do units responding to the same stimulus in naive animals (Miller *et al.*, 1972). Thus differences between studies may arise through differences in the state of arousal of the animals.

A strong response to a natural call does not necessarily mean that the unit has 'extracted' this feature and codes for it. It is probable that other stimuli may also evoke strong responses, including some unrelated to vocalization. One test of the specificity of the call would be to present the stimulus backwards in time. In this way the total spectral composition would be unaltered but the temporal sequence of spectral elements would be completely different. If the unit still responded well to the reversed call, the hypothesis that the unit is specific for the natural call would be falsified. In such cases a thorough knowledge of the response area of the unit to pure tones might explain the effectiveness of the call.

Most studies of the responses of auditory cortical units to species-specific vocalizations have been carried out on squirrel monkeys, New World primates whose vocalizations have been well described both in spectral and temporal composition, and behavioural significance (Winter *et al.*, 1966). Although the auditory cortical fields of this otherwise well-studied species were not mapped with a fine-grain microelectrode analysis in these studies, it is likely, on the basis of a recent analysis of tuning properties of auditory cortex units of the squirrel monkey (Pelleg-Toiba and Wollberg, 1989), that most of the units were isolated in the AI.

The study of Funkenstein and Winter (1973) posed the question of the similarity between responses of units to simple stimuli in the auditory cortex of cats and squirrel monkeys (here, the superior temporal plane and superior temporal gyrus are ventral to the Sylvian fissure). Their records show a high level of spontaneous activity for many units, with inhibitory pauses common to stimuli of any type, all presented from a free-field speaker directly in front of the unanaesthetized animal. A large proportion (71%) of units responded to pure tone stimuli and similar proportions (44 and 41%) of units responded to FM stimuli and calls. It is difficult to relate the response areas measured in other, parametric studies and those of Funkenstein and Winter who carried out 'frequency response scans' at 70 dB SPL. However, they concluded that the responses of units in the auditory cortex of squirrel monkeys did not differ substantially from those of cats (Evans and Whitfield, 1964). It is difficult to see how any other conclusion could be reached, given the diversity of response patterns found in the auditory cortex of both (unanaesthetized) species and the problems of statistical evaluation when different methods and classifications are used.

More pertinent to the present review is the subsequent paper, by Winter and Funkenstein (1973), and that of Newman and Wollberg (1973). Both sought evidence for feature extraction – *viz.* the appearance of responses specific for a particular vocalization, that could not be explained in terms of pure tone response areas. They used 12 vocalizations drawn from the complete repertoire of calls made by the species, but neither the precise nature of the call nor the varieties of response pattern need concern us here. An important observation of both studies was that the response of a unit to pure tones was not a reliable predictor of a response to a natural call, even when the predominant spectral energy of the call fell within that unit's excitatory response range. In some cases the spectral composition of the call fell completely outside the frequency range, yet both were effective. In other cases (7%; Winter and Funkenstein, 1973) units responded to vocalizations but to no other stimulus tested.

There are differences in the two studies in regard to the selectivity to a particular call. For example, Newman and Wollberg (1973) found that a majority (89%) of units responded to more than half of the vocal stimuli, while 41% of those in the study of Winter and Funkenstein were only responsive to one call. However, these studies generally agree that units in the auditory cortex of squirrel monkeys discharge much more powerfully to natural calls than to pure tones and their findings support a concept of selective integration at the cortical level of spectral, intensity and temporal features of an acoustic stimulus.

It would be of great value to know how this behaviour relates to the response area classification of Phillips *et al.* (1985b), which includes both excitatory and inhibitory frequency–intensity domains. Some progress in this direction has been made by Tomlinson and Schwarz (1988), who have been investigating the free-field coding of complex sounds by units in the auditory cortex of alert monkeys after first determining the responses of the units to pure tone and noise stimuli. They find that a 'population of units in the alert auditory cortex can resolve components of a complex tone almost as well as humans in psychophysical experiments'. The classification of Phillips *et al.*, derived from contralateral stimulation, will probably need to be extended for the binaural conditions that prevail in such free-field studies as those of Tomlinson and Schwarz (1988).

It is of interest, at this point, to compare the findings for squirrel monkeys with those for narcotized but unanaesthetized cats, described in a study by Sovijärvi (1975). This study also used free-field stimulation from a source directly in front of the cat. Units were isolated in penetrations that are likely to include sites in the A and AI. Lability and habituation were commonly observed, even though the repetition rate of the 4 s duration stimulus was 1 per 8.5 s, so it would appear that lability is not the province only of unanaesthetized animals. Pure tone stimuli were 80–85 dB so tuning characteristics are not comparable to those of parametric studies.

Units were examined with both pure tones and a variety of taped complex sounds, including cat vocalizations, songbird calls, lemming vocalizations and guinea pig calls. Thirty-two per cent of units responded to complex sounds in ways that were not predictable from their response patterns to selected pure tones. The method of testing for pure tone responsiveness was very restricted in the intensive domain and it is not possible to relate these patterns to the total response area for each unit. However, it is again clear for cats that a proportion (17% in this study) of units in the auditory cortex respond to certain complex sounds but not to any of the pure tones used.

Creutzfeldt *et al.* (1980) compared the responses of cortical and MG units in unanaesthetized guinea pigs to pure tones, both steady and AM or FM, and to natural calls. They observed that cortical neurones responded more phasically to pure tones than did units in the MG, and cortical units were unable to follow the high rates of AM and FM that were effective in the MG. Similarly, in cortical cells, repetitive elements of a call were ineffective if their repetition rate was too high, and they were unable to respond to high modulation frequencies of the fundamental frequency of the call. Creutzfeldt *et al.* further showed that transients were an important stimulus for auditory cortical units and the responses to these and to natural stimuli varied with the intensity of the call. They thus demonstrated that there is a transformation of information about complex sounds between thalamus and cortex.

Human speech has been used as a stimulus for auditory cortical units in unanaesthetized monkeys (Steinschneider *et al.*, 1982) although it is unclear what significance human speech has for monkeys. No information about the spectral characteristics of units to pure tones, steady or modulated, was given in this study and, while clicks were used to distinguish cortical cells and thalamocortical fibres, it is not clear which auditory field was studied. They report a number of findings on the effects of synthetic syllables on cortical responses, including the importance of timing of the onset of the voice and phase locking of formant transitions of the syllables.

Finally it is of interest to note that the technique of magnetoencephalography, where the magnetic fields produced by electrical activity of neurones are measured, can be used non-invasively to assess neural activity in normal human subjects. In the present context, Kaukoranta *et al.* (1987) have been able to map the neural activity evoked by different auditory stimuli over the temporal lobe. They have found that, with human speech, vowel onsets after voiceless fricative consonants evoke a distinct peak of activity in the supratemporal auditory cortex. Using the same recording technique in humans, Hari and Makela (1988) have provided evidence that the masking effects of modulated sounds on human pure tone burst perception have a central, rather than a peripheral, origin. It seems likely that non-invasive techniques such as this will help to relate animal physiology to human perception of communication sounds.

CONCLUSIONS

Studies of coding of acoustic stimuli by cortical neurones fall into two groups – those with a high degree of parametric control but usually in anaesthetized animals stimulated by dichotic stimuli, and those in unanaesthetized, sometimes alert, animals where precision is sacrificed for 'normalcy'. It is clearly important in future studies that every effort be made for each unit to characterize its tone frequency–intensity response areas, location in an organized field and the histology of the recording site before proceeding with the analysis of responses to natural stimuli. It is important to know how well units in the cortex respond to vocalizations but it is probably more important to know what mechanisms underlie these responses. The study of the coding of biologically relevant sounds by units in the auditory cortex remains in its infancy.

The concept that emerges from parametric single-unit studies is one of fixed properties suggestive of a fixed purpose. However, studies that examine the interactions between adjacent neurones reveal that neurones do not act alone but as parts of ensembles with changing compositions. The use of multiple electrodes to record simultaneously the activity of many nearby units in an individual animal has shown that the interactions within a group of neurones depend on the immediate past history of stimulation of those neurones. Espinosa and Gerstein (1988) postulate that a neurone 'participates from time to time in different neuronal assemblies showing a high degree of modifiability in the assembly'.

8 *Binaural physiology of the primary auditory cortex*

Normal hearing involves the integration of neural input from both ears. The presence of two separated acoustic receivers facilitates the discrimination of both the azimuthal angle of a sound source (where?) and the spectral-intensive characteristics of that source (what?). The latter discrimination is aided, particularly for low-intensity sounds, by the spectral and amplitude transformations by the pinnae.

As discussed in Chapter 3, the first major site for binaural integration is the superior olivary complex, and at the midbrain and beyond, more than two thirds of units in the auditory pathway receive input from both ears. Phillips and Gates (1982) suggest that all auditory cortical units may be binaurally influenced with appropriate stimulus conditions. The forms of binaural interaction observed at the cortical level are qualitatively the same as those seen at lower levels (Brugge *et al.*, 1969).

There has been an emphasis, in studies of binaural physiology of the auditory cortex, on the neural encoding of binaural cues necessary for sound localization – the 'where' question. Clinical studies, such as that of Sanchez-Longo and Forster (1958), report that patients having cortical lesions of the temporal lobe experience difficulty in localizing sounds contralateral to the lesioned side. Such clinical data, and a larger volume of animal experimental data (see Phillips and Gates, 1982, for review) suggest a hypothesis that the auditory cortex is necessary for normal sound localization behaviour. This hypothesis has been confirmed by Jenkins and Merzenich (1984) in their classic mapping-lesion-behavioural study.

I will look at the binaural physiology of the auditory cortex by first reviewing the response properties of units to binaural stimuli presented dichotically and in the free field. The question of the organization of response properties across the auditory cortex will then be considered, and an attempt will be made to account for, and predict, sound location perception from this body of experimental data.

THE RESPONSES OF CORTICAL NEURONES TO BINAURAL STIMULI PRESENTED DICHOTICALLY

The responses of many single units in the primary auditory cortex of cats and monkeys to binaural dichotic tonal stimuli are influenced by the relative times of arrival and intensities of sounds at each ear (Brugge *et al.*, 1969; Brugge and Merzenich, 1973; Kitzes *et al.*, 1980; Phillips and Irvine, 1983). The effects of variations in the interaural time difference (ITD) can take two forms. First, the probability of firing may be dependent in a periodic fashion on the ITD, with the period being approximately that of the stimulating pure tone (Brugge *et al.*, 1969). Units sensitive periodically to ITD behave as though they receive both excitatory and inhibitory inputs from each ear that are locked in phase to the pure tone. Thus for the unit illustrated in Figure 8.1, maximum firing probability is observed at the contralateral delay of 100 μs and this peak value is greater than the sum of the individual contralateral and ipsilateral responses (i.e. facilitation has occurred). At unfavourable delays (e.g. 200 μs ipsilateral delay) the firing rate is below that for contralateral stimuli, the effects of which must therefore have been reduced by the ipsilateral input. Given the cyclic nature of the spike count–delay curve (Figure 8.1a), one must conclude that ipsilateral stimuli, at least, evoke alternating phase-locked excitatory and inhibitory input to this unit. The changes in first-spike latency (Figure 8.1b) and the variation in latency (Figure 8.1c) probably reflect, rather than are causative of, the variation in firing probability.

Periodic ITD effects have been observed in units with CFs below 2.5 kHz (Brugge *et al.*, 1969). Although an upper frequency limit for this phenomenon has not been determined, it certainly appears to be a common form of binaural interaction for low-CF units at all levels of the auditory pathway above the superior olivary complex (e.g. Irvine, 1986). Many units show maximum responses at ITDs that would occur in nature (for cats, 0–400 μs); such units are likely to have a role in the localization of low-frequency sounds.

A second manifestation of an ITD effect is explicable in terms of latency differences between single (non-phase-locked) synaptic actions evoked by independent stimulation of each ear. In this situation variation in the ITD may lead to the suppression of an excitatory response, a suppression that may be total and maintained for long (several milliseconds) interaural delays (Caird and Klinke, 1983). This interaural arrival-time disparity may be observed at any CF, and need not always be suppressive. For some units, monaural stimulation of either ear may have little effect, but binaural stimulation with close to zero delay leads to a strong response (Kitzes *et al.*, 1980; Phillips and Irvine, 1983).

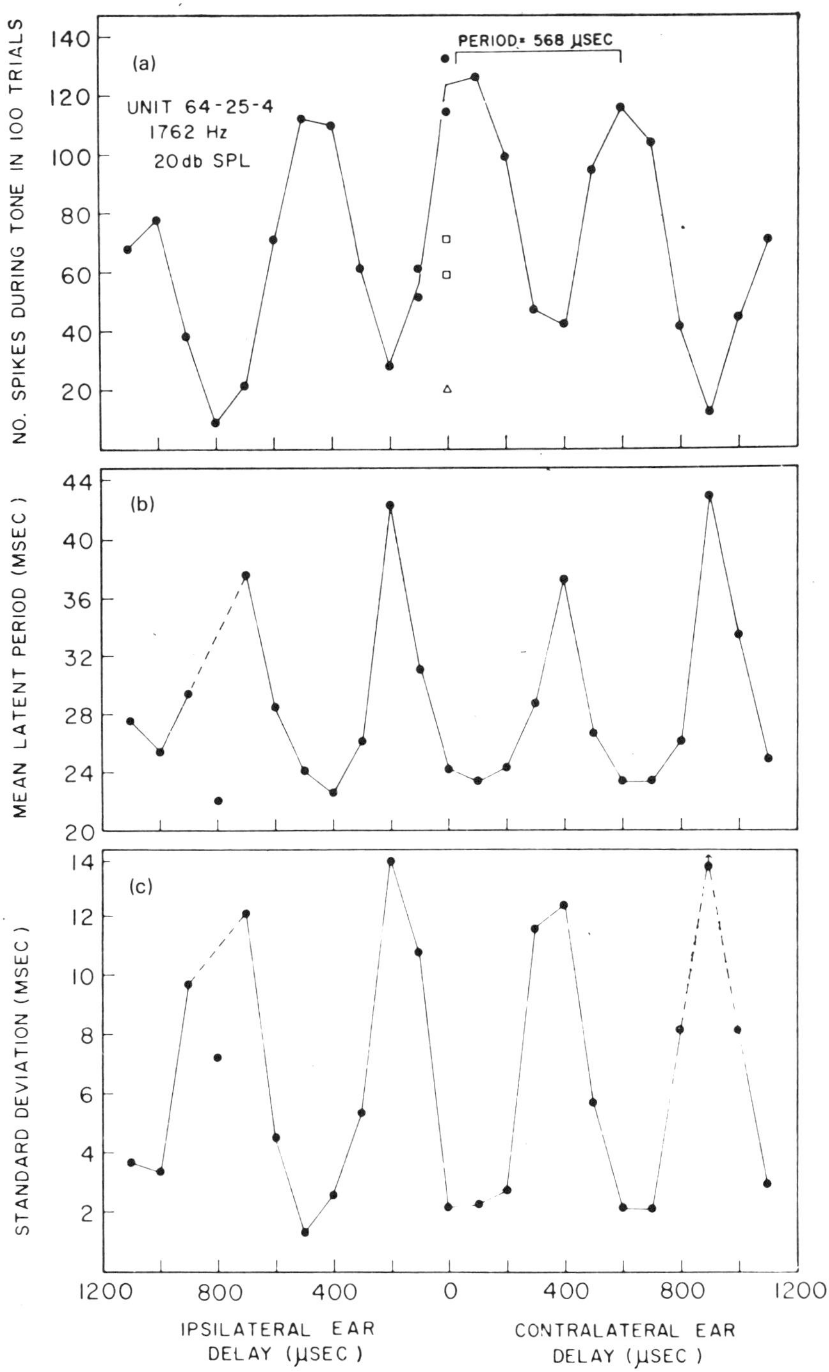
NO. SPIKES DURING TONE IN 100 TRIALS
(a)
UNIT 64-25-4
1762 Hz
20db SPL
PERIOD= 568 µSEC
MEAN LATENT PERIOD (MSEC)
(b)
STANDARD DEVIATION (MSEC)
(c)
IPSILATERAL EAR DELAY (µSEC)
CONTRALATERAL EAR DELAY (µSEC)

The influence of interaural intensity disparities (IIDs) on responses of auditory cortical neurones may also be facilitatory or suppressive (Brugge *et al.*, 1969; Phillips and Irvine, 1981, 1983). The former effect is seen commonly with neurones of the EE type (excitatory input from each ear) while suppressive effects are observed in EI neurones (excitatory–inhibitory interactions). A higher proportion of neurones are binaurally influenced in the AI than in the IC (Hall and Goldstein, 1968; Phillips and Irvine, 1983; Roth *et al.*, 1978; Semple and Aitkin, 1979). This difference could be the result of convergence within the neuronal population projecting to the cortex, the selective projection of monaural inferior collicular neurones outside the ascending auditory pathway, and the influence of higher-level crossed auditory projections (commissure of the inferior colliculus and corpus callosum).

That all three of the above factors may play a part in determining colliculo-cortical binaural transformations is suggested by the emergence at the AI of new types of binaural interaction not observed at the inferior collicular level. In the IC, all studied EI cells are excited by contralateral stimuli and suppressed by ipsilateral input (Geisler *et al.*, 1969; Roth *et al.*, 1978; Semple and Aitkin, 1979; Semple and Kitzes, 1987a). This property can in turn be traced back to the contralateral projection of the high-CF part of the lateral superior olivary nucleus, where EI behaviour first arises (see Chapter 3).

Excitation of high-CF inferior collicular neurones with EI properties occurs when the contralateral intensity exceeds or equals that at the ipsilateral ear, whereas suppression is exhibited when the ipsilateral exceeds the contralateral intensity. In other words, only the contralateral sound field is represented by the discharges of high-CF neurones in this nucleus.

In the AI it has long been known that ipsilateral stimuli can evoke greater activity than contralateral stimuli at certain loci (e.g. Woolsey and Walzl, 1942) and may be the only excitatory ear with certain single units (e.g. Phillips and Irvine, 1983). In a recent study (Irvine *et al.*, 1989) we have observed a variety of effects of interaural intensity disparities (Figure 8.2).

Spike count versus IID functions are usually similar in shape for a given unit as a function of intensity (Figure 8.2, upper row) so that, at suprathreshold average binaural intensity levels, it is appropriate to take

Figure 8.1 Curves showing effects of interaural time delay on number of spikes (a), latent period to the first spike (b), and dispersion of the latency values around the mean (c). Closed circles correspond to binaural responses; open squares and triangle in (a) correspond to contralateral and ipsilateral responses, respectively. From Brugge *et al.* (1969).

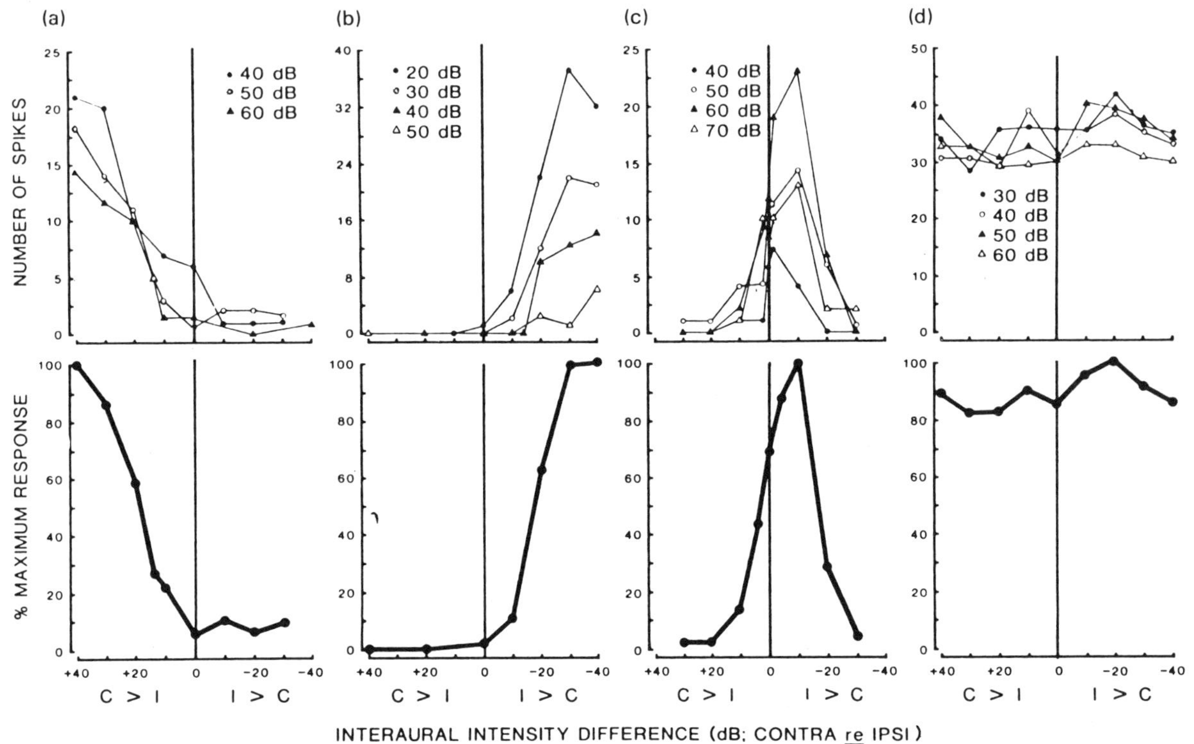

Figure 8.2 Upper: Representative functions relating interaural intensity difference at various average binaural intensities (symbols with dB values), to spike count. Lower: Averages of the functions shown above plotted against percentage of maximum response. (a, b, c) Functions peak in the contra maximum range, ipsi maximum range and near zero respectively. (d) Insensitive to interaural intensity differences.

an average function by computing the arithmetic mean of all intensity functions for a given unit (Figure 8.2, lower). Of these mean functions, 45% of high-CF neurones in the AI are sensitive to contralateral (Figure 8.2a), and 16% to ipsilateral, interaural intensity difference ranges (Figure 8.2b). In other words, neuronal activity in some parts of the AI codes events occurring in the ipsilateral sound field.

The type of cell unresponsive to monaural stimuli but responsive to binaural stimuli (predominantly binaural, or PB cells) (Kitzes *et al.*, 1980; Phillips and Irvine, 1983) forms 9% (Hall and Goldstein, 1968), 12% (Phillips and Irvine, 1983) and 17% (Irvine *et al.*, 1989) of the sample recorded in the AI. These neurones respond most strongly when the interaural intensity difference is near zero (Figure 8.2c), and would thus be most sensitive to sounds located near the median plane. Such neurones are rarely observed in the IC.

It is interesting to speculate how ipsilateral PB cells might occur at the cortical level yet be absent in the IC. The ipsilateral excitation may derive from the fact that 10–20% of afferent input to the medial geniculate (MG) originates from the contralateral IC via the commissure of the IC (Aitkin and Phillips, 1984; Moore and Goldberg, 1963). This projection would effectively contribute a 10–20% increase in ipsilateral excitation to that ascending through the MG to the AI. If this is the case, neurones with ipsilateral sound field preferences should be identifiable in the MG, as in the AI. In fact two of 17 units studied in the MG by Ivarsson *et al.* (1988) were excited at ipsilateral IIDs.

The corpus callosum of cats distributes mostly to areas containing EE cells and appears not to interconnect EI regions (Imig and Brugge, 1978). One feature that might be enhanced by this commissural connectivity pattern is the representation of the median plane, just as central vision is amplified by callosal connections of the visual cortex. The PB cells of the AI would show clear preferences for frontal locations and their increased numbers over those observed in the IC may be a product of interhemispheric connections. Against this argument, Moore *et al.* (1974) have shown that lesions of the corpus callosum do not influence sound localization behaviour in cats; according to the speculation above, callosal lesions should influence localization in the median plane but not the left and right periphery. Furthermore, analogous cells to the PB class have been described in the MG by Ivarsson *et al.* (1988). Their appearance in numbers at the thalamocortical level could be due to convergence and integration within the ascending auditory pathway of each side, rather than be a product of commissural connections.

Semple and Kitzes (1987a,b) have analysed the responses of units in the superior olive, inferior colliculus and the auditory cortex to IID, average binaural intensity (ABI) and combinations of these intensity parameters. A type of high-CF neurone, referred to as 'TWIN', shows non-monotonic changes in spike count as a function of increases in either the IID or the average binaural intensity. As a consequence such a unit fires maximally at a certain IID and ABI, and would be a potential candidate for coding azimuthal angle at a particular intensity. Of interest to the present discussion is that while no TWIN units are observed in the superior olive, they form about 10% of the population in the IC and 47% in the AI. Semple and Kitzes (1987b) consider that such joint sensitivity to IID and ABI is likely to be a product of hierarchical processing of information.

RESPONSES TO FREE-FIELD STIMULI

Dichotic stimulation allows independent manipulation of ITD and IID, and has revealed the potent influences of these cues on certain binaural neurones. However, this is an artificial situation, since a single sound source in space will generate both an ITD and an IID for each position. Neither of these cues is a simple linear function of sound source azimuth (Irvine, 1987; Roth *et al.*, 1980) so that each speaker position will be associated with a set of ITD/IIDs that may be unique. In addition, sound reaches the two inner ears after a spectral and amplitude transformation by the pinnae. The spectrum-versus-amplitude composition of a complex sound reaching the eardrum will differ from that leaving the source, and will also differ according to azimuthal angle (Shaw, 1982; Carlile and Pettigrew, 1987). As a consequence of these factors, it is not possible at the outset to deduce free-field directional characteristics from data derived from dichotic studies for anything other than pure tone stimuli.

The pinna's directional amplification may indeed be a potent influence at low sound intensities. It has been shown that, for cats, up to 30 dB amplification is possible at certain frequencies (e.g. 3 kHz) when sounds fall on the acoustical axis of the pinna (Phillips *et al.*, 1982). However, the discharge rate of an individual unit responding to this 3 kHz sound can only differ between on-axis and off-axis locations if it has not saturated at the off-axis position. If it had saturated it would therefore be responding maximally to 3 kHz tones at both on- and off-axis positions. Recall (Chapter 7) that the dynamic range of units of the AI, at least under dichotic contralateral conditions, is about 30 dB. Sounds from threshold to 30 dB above threshold could cause responses from a unit if the sound source is on-axis but may not if the source is off-axis. At intensities higher than about 20–30 dB above threshold the firing rate for on-axis positions would not change but would become measurable for off-axis positions. At 60 dB above threshold there would be nothing in the firing rate of that unit to distinguish on- and off-axis 3 kHz sound source locations. It should be noted that for most frequencies the pinna amplification is less than this, being closer to 10–20 dB (Phillips *et al.*, 1982).

Now, owing to the diffracting effect of the head, sounds of high frequency 10–20 dB above threshold, emanating from a sound source placed on the acoustic axis of one pinna, will be of too low intensity to stimulate the opposite ear. Thus essentially monaural cues given by the pinna provide at low intensities a potential mechanism for sound localization that will be supplanted by binaural cues of ITD and IID at higher intensities.

In this regard, the study by Middlebrooks and Pettigrew (1981) of units in the AI responding to free-field sound is of interest. These authors used a CF stimulus intensity 10 dB above threshold and measured the spatial region within which that stimulus evoked excitation – the spatial receptive field (RF) for that unit. They were able to group RFs into three classes – omnidirectional, hemifield and axial. Omnidirectional units responded to stimuli placed anywhere in front of the cat, and for the majority the most effective stimulus was located near the median plane. Hemifield RFs occupied most of the contralateral sound hemifield and had sharp borders within 20° of the midline. Axial units had RFs located in line with the acoustic axis of the contralateral pinna. The boundaries of the RFs were not systematically tested at higher intensities, but the authors indicate that the boundaries were fixed for some units up to 30 dB above threshold while for others the boundaries expanded. Most axial units had CFs above 12 kHz, most hemifield CFs fell below 12 kHz, while omnidirectional units were spread equally above and below this value.

Let us interpret these observations in the light of the above discussion of pinna directionality, head shadow and dichotic classification schemes. First, the omnidirectional class presumably includes some EE and PB cells, particularly the latter since the most effective stimuli were near the midline. Hemifield units include EI cells and those EE cells for which the ipsilateral threshold is in excess of 10 dB above the contralateral threshold. Axial units, however, are difficult to characterize since at the frequencies and intensities employed they are dominated by the passive acoustics of the contralateral pinna. The fact that there are many axial units at CFs above 12 kHz may be due to the considerable diffraction by the head and pinna, at these high frequencies, for all positions other than that in line with the acoustic axis of the contralateral pinna. One must conclude that, for many units in the study of Middlebrooks and Pettigrew, only monaural and not binaural stimulation was achieved at these lower intensities.

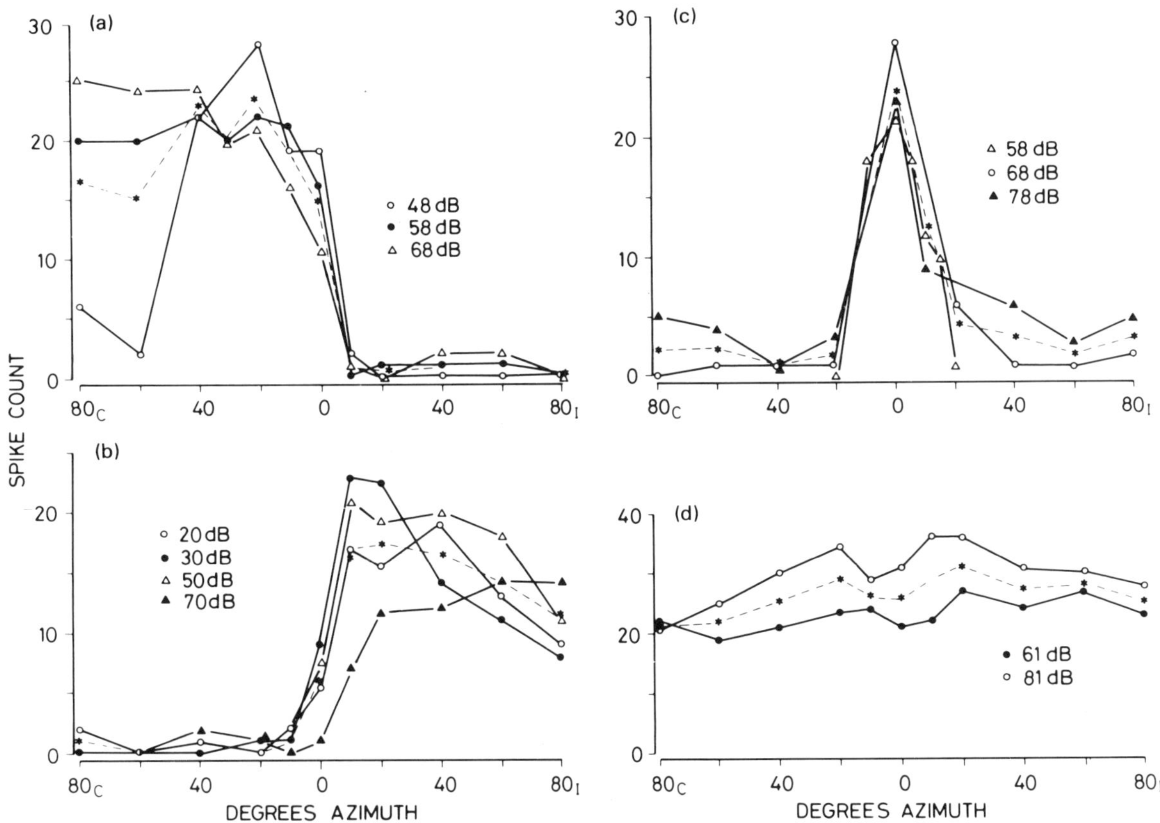

(a)
48 dB
58 dB
68 dB
(b)
20 dB
30 dB
50 dB
70 dB
(c)
58 dB
68 dB
78 dB
(d)
61 dB
81 dB
SPIKE COUNT
DEGREES AZIMUTH

A very different picture of directionality is provided when sound stimuli are in excess of 20 dB above threshold, and firing rate, rather than RF boundary, is the response measure. Aitkin *et al.* (1988b) define four classes of unit based on the shapes of their 'mean azimuth functions', plots of firing rate against azimuthal angle averaged for several different intensities in excess of 20 dB above threshold (Figure 8.3). This classification is based on an earlier study by Eisenman (1974). 'Contra-field' units (Figure 8.3a) (46% of the population) respond to stimuli located mostly in the contralateral hemifield, and exhibit a border of varying sharpness within 20° of the median plane. 'Ipsi-field' units (Figure 8.3b) (17% of the population) have identical properties except for the effective hemifield being ipsilateral. 'Central' units (Figure 8.3c) (11%) respond to positions in the median plane and tend to become unresponsive at positions more than 20° away from the midline, and omnidirectional units (Figure 8.3d) account for 20% of the population. These classes overlap and may be part of a continuum; as a result some units (6%) cannot be assigned to any of these classes. These proportions are similar to those for the small number of units studied by Eisenman (1974), differing in the number of omnidirectional ('full field', 28%) and contra-field (33%) units observed by that author.

There is a striking correlation, with dichotic and free-field class components in the parallel studies of Irvine *et al.* (1989), between contra-field and contra-IID (46% and 45%), ipsi-field and ipsi-IID (17% and 16%) and central and PB (11% and 17%). This correlation strongly suggests that, for high-CF neurones of the AI, the behaviour of a unit in the free field can, at intensities well above threshold, be predicted from its response to interaural intensity disparities, and *vice versa*. Thus, at suprathreshold intensity levels of pure tones, the directional effects of the pinna appear to have little importance.

Sound localization will often require judgements to be made of a moving sound source or by a moving listener. Sovijärvi and Hyvärinen (1974) examined the responses of auditory cortical cells to both moving and stationary stimuli. Only a small number of units was examined, and it is not always clear how the sound stimuli used related to the CF of each unit or its threshold. However, 8/25 units studied gave responses to moving sounds that depended on the direction of movement. Movement in a favourable direction evoked sustained excitation, and in the opposite direction, suppression or no response. For half of this sample, the

Figure 8.3 Representative functions, for four units, (a)–(d), relating the azimuthal location of a sound source in degrees, to spike count. Solid lines and symbols: individual intensities. Dashed lines: average function for that unit. (a), (b), (c), (d): contra-field, ipsi-field, central and omnidirectional units, respectively.

responses of the cell to stationary sounds of the same spectral content at different points along the arc of movement were different to those during movement. This early study needs to be replicated parametrically because it confirms observations, made in the MG by Altman *et al.* (1970), that a population of neurones in the thalamocortical auditory system is specifically concerned with the detection of a moving sound source.

ORGANIZATION OF BINAURAL NEURONES ACROSS THE AUDITORY CORTEX

Since the studies of Knudsen and Konishi (1978) of the representation of auditory space in the auditory midbrain of the barn owl, a number of experimenters have sought to find evidence for 'space maps' in the mammalian auditory pathway. Teleological considerations would suggest that maps of auditory space may have evolved in predatory species that primarily localize and capture by means of the sounds made by their prey. In this respect the owl may be highly specialized, since predatory species such as humans, cats and diurnal hawks probably use their visual systems for target location; the auditory system being an adjunct for alerting and arousing the species. Nonetheless, evidence is accumulating for representations of auditory space, or more particularly azimuthal angle, in a number of regions (Aitkin *et al.*, 1985, 1988b; Imig *et al.*, 1987; King and Palmer, 1983; Middlebrooks and Knudsen, 1984; Wise and Irvine, 1985). A basis for such a representation may be found in the binaural columns or bands of the AI of cats found in dichotic mapping experiments (Imig and Adrian, 1977; Middlebrooks *et al.*, 1980).

Imig and Adrian (1977) used a qualitative classification to characterize neuronal excitatory responses to monaural and binaural stimuli at CF. Monaural responses were contralateral or ipsilateral dominant, or equidominant; binaural responses were summative (greater than either monaural response alone), suppressive (less than the dominant ear alone) or occlusive (equal to the dominant ear response). Unfortunately this scheme is difficult to relate to the conventional classification schemes of binaural interaction, and it is not clear if the class of a unit changed for different stimulus intensities.

Imig and Adrian recorded the binaural interactions of sequentially isolated units and found that cells, in columns of variable widths extending from the white matter to the surface, displayed similar binaural interactions. Penetrations oblique to the cortical surface recorded groups of cells with alternating binaural interactions. Within suppression columns the contralateral ear was dominant, but within summation columns the ear dominance varied.

They also mapped the distributions of interactions and dominances across the AI by making multiple microelectrode penetrations and recording from two or three points in each penetration. The resultant maps suggest several summation columns and at least one suppression column oriented as strips of cortex from rostral to caudal – i.e. in the direction of tonotopic organization (Figure 8.4). Thus they provided evidence for the AI of cats having cells arranged in strips with similar CFs and differing binaural characteristics in one direction, and similar binaural characteristics but changing CF in the orthogonal direction. In so doing they confirmed Tunturi's (1952) observations made in the auditory cortex of dogs some 25 years earlier.

Middlebrooks *et al.* (1980) extended these observations by making very wide and detailed maps of binaural interaction classes across the AI. They used the more conventional EE/EI nomenclature but defined the class of a unit near its excitatory threshold. There is no certainty that this class would be maintained for a given unit at higher sound levels, and it is known (e.g. Brugge *et al.*, 1969) that ipsilateral stimuli may have both excitatory and inhibitory influences. However, it seems likely that the 'summative' class of Imig and Adrian and the EE class of Middlebrooks *et al.* would be largely equivalent, as would the 'suppressive' and EI groups.

In several experiments, four or five very long tangential penetrations, recording mainly from the middle layers, detected the presence of a central EI band flanked by two EE bands, confirming the observations of Imig and Adrian. Evidence for additional EI zones, one dorsal and one ventral to the EE bands, was also put forward, giving a picture of the AI with binaural bands each with the same interaction class, intersecting multiple isofrequency contours. Middlebrooks *et al.* (1980) speculated, on the basis of these multiple cochlear representations, that the AI may contain a number of 'independent processing units', each concerned with a different aspect of auditory perception.

Both of the above studies employed stimulus intensities close to CF threshold. Reale and Kettner (1986) used a wider intensity range in their study of the distribution of binaural interaction classes across the AI. Their results generally agree with those of Imig and Adrian, and of Middlebrooks *et al.*, with the additional recognition that some units have mixed binaural interactions that do not fit in with summation or suppression classes.

These studies suggest that binaural input is organized differentially across the AI of cats. Does this mean that the location of a sound source is the feature represented in this dimension? As stated at the beginning of this chapter, sound localization is only one outcome of binaural hearing. Neurones receiving excitatory input from each ear (EE) could, for

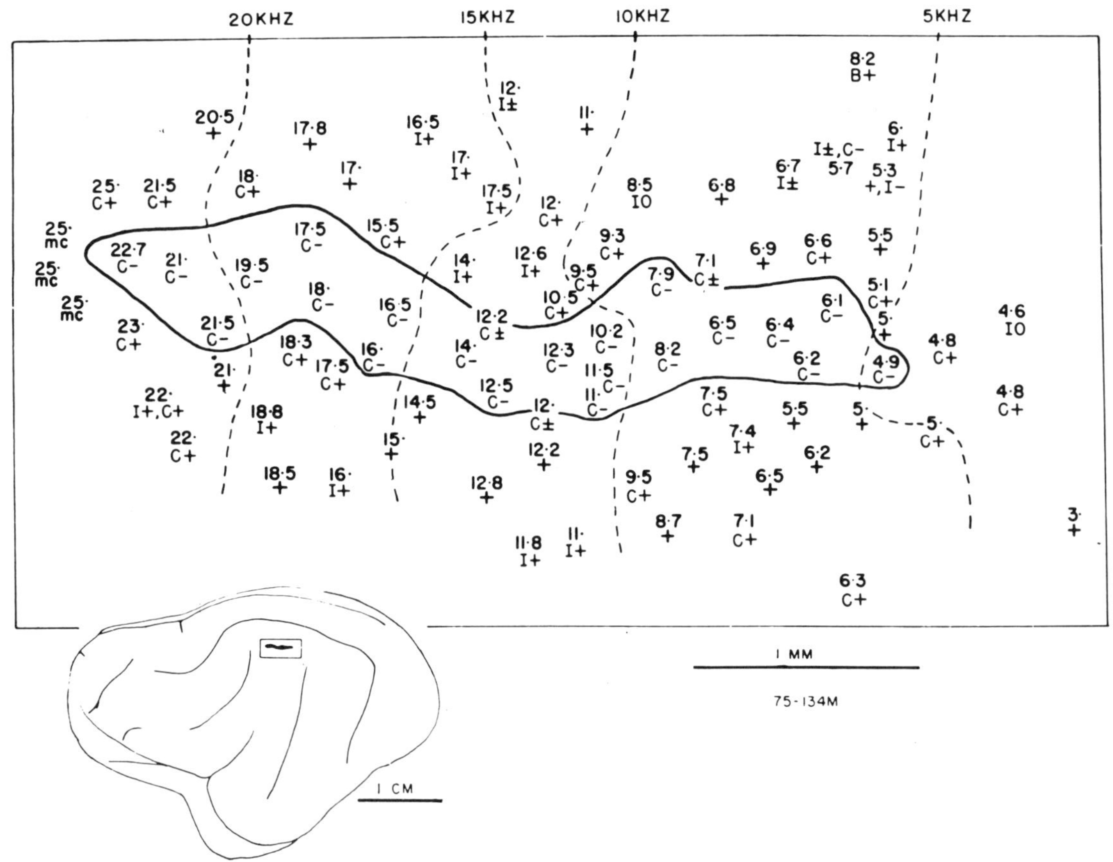
20KHZ
15KHZ
10KHZ
5KHZ
1 MM
75-134M
1 CM

example, be concerned with the recognition of the nature, rather than the location, of a stimulus. Middlebrooks and Pettigrew (1981), in an effort to answer this question by using free-field acoustic stimuli, were able to show alternating receptive field classes in long penetrations tangential to the AI. However, as stated previously, it is unlikely that they were very often mapping the outcome of binaural interaction although it is possible that their 'omnidirectional' unit class is comparable to those of the dichotic stimulation experiments, viz. the summative or EE class.

Measurements of azimuthal sensitivity at higher sound pressure levels in the AI of cats by Imig *et al.* (1987) and by Aitkin *et al.* (1988b; unpublished) reveal that the most effective azimuthal location may vary systematically as a function of location along an isofrequency strip. This seems to be manifested in several ways that are not exclusive. First, there are changes across free-field response classes, so that, for example, contra-field units may grade to central and then to ipsi-field classes, corresponding to a sound source being moved from one side, through straight ahead to the other. A second trend is for the azimuthal range giving an 80%-of-maximum response to shift from the contralateral pole to the midline and then back again. This corresponds to a mapping of the contralateral azimuthal plane. A third observation is for omnidirectional units to be scattered throughout a sequence of directional units, rather than forming a dense grouping. Whatever function these units have, they do not mask a given trend. Finally, there may be more than one sequence in a given long penetration.

Variation is a constant feature: azimuthal mapping trends vary from penetration to penetration, and from animal to animal. This example of maps varying across individuals is more extreme than the variation of the frequency map (Merzenich *et al.*, 1975) and is concordant with the concept of 'personalized' maps in the auditory cortex (Suga *et al.*, 1987). This variability makes a correlation with binaural bands (which probably also vary across individuals) difficult, but a number of possibilities exist.

Figure 8.4 Topographic distribution of binaural interactions, best frequencies and aural dominances of neurone responses. Isofrequency contours are indicated by dotted lines; plus, summation at one depth in penetration, suppression at another. The hemisphere drawing shows the portion of auditory cortex which was mapped. Solid lines separate suppression columns from summation columns. Interrupted lines partition the summation columns into areas with similar aural dominance. Best frequencies are indicated in kHz. If neurone response characteristics changed as a function of depth in a penetration, both responses are shown on the map separated by a comma; plus, summation; minus, suppression; O, occlusion; C, contralateral dominance; I, ipsilateral dominance; B, binaural response, no monaural response; + (in absence of a C, I or B), equidominance; MC, monaural contralateral response. From Imig and Adrian (1977).

First, contralateral azimuths may be mapped in EI bands, and central/ipsi azimuths in EE bands. Secondly, cluster responses to stimulation of either ear may be composed of the responses of ipsi-field, contra-field and omnidirectional units mixed together; on this basis 'EE bands' may contain heterogeneous free-field cell classes. Whatever the representation, it would seem to be repeated across the surface of the AI at right angles to its tonotopic axis.

THE AI AND SOUND LOCALIZATION BEHAVIOUR

In the earliest experiments relating the cerebral cortex to sound evoked behaviour, Ferrier (1876) showed that lesions of the temporal lobe of a number of species led to a neglect (in terms of head turning or oriented pinna movements) of sounds occurring contralateral to the lesion. This correlated with Ferrier's observations that electrical stimulation of this temporal region could cause orienting movements of the contralateral pinna.

Clinical observations in humans with unilateral temporal lobe damage have often been more anecdotal in relation to sound localization. Two studies directed towards this question have produced rather different results, but this could be due to the methodologies employed. Sanchez-Longo and Forster (1958) employed a free-field localization task to test patients with unilateral temporal lobe lesions, and observed performance deficits in most of these patients for sounds originating from the contralateral hemifield. Walsh (1957) employed a dichotic listening task and varied ITD in order to see if patients could lateralize the sound to the left or right ear. The fact that they lateralized sounds with relatively normal accuracy could mean that the areas of damaged cortex were not necessary for the task, or it could mean that some other perceptual change (e.g. loudness) gave a clue to the subject. 'Left' versus 'right' is a rather different question to 'where in space', for in the latter case no choice is offered and the possibilities are infinite.

A problem with the incorporation of human clinical evidence into a model of the AI and sound localization is that the extent and degree of damage to the AI is difficult to specify for human patients, and other areas are usually damaged in addition to the AI. Such a problem can be avoided with animal experiments, but then the behavioural paradigm can be complex and the results subject to interpretation. For example, does the inability of an animal to go to the correct sound source location to receive a food reward mean that it cannot perceive the correct azimuthal angle, or might not the problem be one of acousticomotor integration? On the other hand, does a correct choice in a left versus right discrimination mean a correct localization judgement, or could some other cue be involved?

Jenkins and Merzenich (1984) designed a behavioural paradigm in which some of these confounding artefacts were eliminated. Cats were trained to localize noise bursts and some pure tones to one of seven speaker/food boxes in a semicircle of radius 89 cm (speaker separation 30°). When trained, the cats were anaesthetized and the AI mapped using microelectrode procedures. Cortical lesions were made by coagulating all cortical blood vessels entering the selected region, usually an 'iso-frequency' strip of cortex. Aseptic conditions were used throughout, and after the animals recovered from the operation they were retested in the behavioural paradigm.

Striking deficits were found for contralaterally located short duration (40 ms) tone bursts of specific frequencies. These frequencies lay within those that had been represented in the strip of cortex prior to the lesion. At these frequencies, ipsilateral deficits were quite small and within the range of variability seen in normal cats. For the latter, incorrect choices were usually of adjacent speakers, and accuracy improved with continuing trials. For the lesioned animals, incorrect choices could involve speakers remote from that used as a sound source, and the deficit was permanent. However, the deficits observed in lesioned animals were only for short-duration sounds, and performance was near normal when durations of 200 ms were used. Jenkins and Merzenich considered that cats were able to use orienting movements and scanning movements of the pinnae to aid their localization ability with longer tones. If this is so, the deficits to short tones could be considered as 'perceptual' rather than acousticomotor since the animals could make the correct response if the stimulus was adequate. The authors further note that the animals 'heard' these short sounds very clearly because they moved quickly, although inaccurately, to the supposed sound source. Therefore there was no suggestion of elevated thresholds with these cats.

Jenkins and Merzenich stress that deficits were mainly for contralateral locations, with cats performing at near-normal levels for ipsilateral and median plane locations. This does not tally completely with the physiological observations that some 32% of neurones are most sensitive to sound sources located centrally and ipsilaterally. It is possible that the lesions did not interfere with the ipsilateral or central azimuthal representations. Alternatively, it may be that the populations of neurones in the opposite, undamaged AI were adequate functionally to ensure normal localization even in the absence of those on the lesioned side. In other words, if a certain basic number of auditory cortical neurones is intact, correct localization performance will occur.

These observations show that the AI is necessary for normal

contralateral sound localization in cats. Is the retention of AI, in the absence of other auditory cortical areas, sufficient for normal localization? Jenkins and Merzenich suggest that it is, on the basis of an experiment in which almost all auditory cortex, *other than* an isofrequency strip in the AI, was removed. The strip extended from the most dorsal to ventral limits of the AI and contained neurones responding best over a range of 3–11 kHz. The cat was able to localize this range of frequencies accurately in the contralateral hemifield but localized other frequencies very poorly. The authors conclude 'the normal localization performance at these frequencies strongly suggests that AI and its subcortical connections are sufficient to support normal azimuthal sound localization behaviour' (p. 837).

How is sound-localizing behaviour effected? The specific pathways, descending from the cortex, that direct motor responses for sound orientation and localization have not yet been dissected, but from the above reasoning they would, in cats, appear to arise in the AI and not a secondary field. This would seem to exclude a direct link with the superior colliculus, projections to which arise in the AII (Chapter 5). One likely pathway from the AI is that terminating in the basal ganglia; another possible connection may involve the IC. It is highly probable that these long-latency, non-reflexive behaviours are the outcome of a variety of interactions from diverse pathways, all of which involve the auditory cortex.

9 The development of the auditory cortex

In this chapter I will examine how the neurones of the auditory cortex, and their connections and electrophysiology, develop during gestation and postnatally. There is little doubt that the processes of development are not restricted to gestation and the period immediately after birth, but may continue throughout a significant segment of life. For this reason I will also examine the question of the plasticity of auditory cortical organization in the adult.

When can an individual of a species be considered as mature? One marker could be the time of onset of reproductive capacity, but maturation certainly continues beyond this point with humans. However, aside from humans, the development of auditory function during adolescence has hardly been studied in any species. For humans it is likely that hearing, *per se*, has reached maturity early in youth, and that subsequent auditory development is that relating to language and communication.

When does hearing begin? Is it at the onset of peripheral auditory function, when information is first transmitted into the brain (e.g. Pujol and Hilding, 1973), or does it begin when output from the auditory system occurs? For convenience, let us take, as one landmark for viewing auditory cortical development, the time of occurrence of a behavioural response to sound. Ehret (1983) has reviewed the literature on startle response and pinna reflex initiation as indices of behavioural responses to sound. Startle responses are whole-body twitches that occur at short latencies, and pinna reflexes are evoked by similar intensities of sound in species with mobile pinnae (see Aitkin, 1986). Equivalent responses occur in humans 16 weeks before birth, guinea pigs 5–6 hours after birth, cats at 5 days postnatal, ferrets (Moore, 1982) and mink at 30 days postnatal and the marsupial opossum at 50 days postnatal. A behavioural response of this sort indicates an active linkage between acoustic input and motor output, but it is very primitive, and more sophisticated behavioural responses, such as orienting reflexes, take longer to develop.

The much earlier figure for behavioural responses of humans is based on two sorts of evidence: averaged evoked responses to sound stimuli recorded from the brain stem of premature babies (e.g. Starr *et al.*, 1977), and foetal movements evoked by vibrotactile stimuli applied to the mother's abdomen (e.g. Birnholz and Benacerraf, 1983). Both approaches have uncertainties. First, premature babies are no longer foetal, and pneumatization of the middle ear will have occurred, possibly triggering or accelerating maturation of the cochlea that normally occurs at term. Secondly, foetal movements evoked by vibrotactile stimuli may be signalled by cutaneous, rather than cochlear, mechanoreceptors. A more appropriate comparison with the animal studies would be the initiation of a startle response or pupillary reflex in human babies. Conversely, a study using radioactive 2-deoxyglucose to measure functional activity in the brain has shown that such activity occurs in the inferior colliculus of foetal guinea pigs exposed to pure tone frequencies ranging from 0.8 to 20 kHz (Horner *et al.*, 1987). This suggests that the 'onset of hearing' in guinea pigs may have to be brought forward from birth to at least 10 days before birth. While the presence of a startle response is a positive sign of sensory input, its absence does not rule out sensory input to the brain. In the foetal sheep, brain stem auditory evoked potentials first appear at around gestational day 117, well before term (147 days) (Cook *et al.*, 1987; Woods *et al.*, 1983).

Whatever the criteria used for assessment, it is apparent that the timing of maturational events may differ considerably from one species to another, and certain processes may occur *in utero* in some eutherians, postnatally in others, and in the pouch in marsupials. In the descriptions of the development of the auditory cortex that follow, the sequence of events is likely to be similar in most species, even though the absolute time relative to 'birth' at which they occur will be different.

MATURATION OF AUDITORY CORTICAL STRUCTURE

The cells of the auditory cortex, like those in all of the neocortex, are generated in the primitive ependymal cell regions around the lateral cerebral ventricles and migrate to their final destinations in the cortical plate. As shown in the classic study of Angevine and Sidman (1961), carried out on mice, the first cells to be established in the cortex are those of layer VI (in mice, 11th gestational day) and the last, layer II (16–17th gestational day). The pattern for the auditory cortex of a marsupial, the brush-tailed possum, is very similar to that for the eutherian mouse, but the period of time over which migrations occur is much greater in the possum (Sanderson and Aitkin, unpublished observations). This inside-out pattern is quite different to the migration sequence in lower auditory nuclei, in which distal cells are first laid down, followed by more proximal cells (Altman and Bayer, 1981; Cooper and Rakic, 1981). Angevine and Sidman (1961) speculate that synaptic contacts may be attained when cells, destined for superficial cortical layers, migrate past those in already established layers. These synaptic contacts might lead to the connective arrangement, characteristic of the neocortex, in which the long axons proceed into the white matter, and long apical dendrites (of pyramidal cells) pass towards the pia.

The establishment of the cortical plate is associated with the growth of axons. Initially a primordial layer can be recognized composed of an outer layer (layer I) and an inner layer (VI) (Marin-Padilla, 1971). As maturation continues and more cells arrive at the cortical plate, layers I and VI develop structural and functional interconnections and begin to receive cortipetal axon collaterals. Marin-Padilla (1971) considers that this primordial neocortical organization resembles the organization of the reptilian neocortex. In cats (60 day gestation period) such a primitive transitory organization appears between 25 and 45 days' gestation, with afferent fibres arriving at the cortical plate at 45 days. From this time on pyramidal cells mature, spines develop and the cortical organization changes from a reptilian type to the typical, layered mammalian neocortical structure.

The growth of auditory thalamocortical fibres has been studied in fixed human foetal tissue using a histochemical method for the enzyme acetylcholinesterase. These fibres emerge from the diencephalon at 10 weeks' gestation, and over the next 10 weeks grow towards and penetrate the auditory cortex (Krmpotic-Nemanic *et al.*, 1980, 1983). The final stage of growth is correlated with the development of cortical lamination, and it would appear that these processes occur somewhat earlier during the gestation period for humans than they do in cats. This might contribute to the earlier overall maturation of the auditory system of humans.

The process of myelination begins during foetal life and continues in most species well beyond birth. In the auditory cortex of dogs, myelin is not detectable until 4 weeks postnatally (Fox, 1968). In cats the white matter thickens greatly from 45 days, due partly to the arrival of the first callosal fibres from the opposite hemisphere (Marin-Padilla, 1971). Postnatally in cats, the latencies of single-unit discharges and evoked potentials decrease at all levels of the auditory pathway with increasing age (Pujol, 1972); latencies of cortical responses diminish from as much as 60 ms to about 10 ms over the first 3 weeks postnatal. Such a latency reduction is likely to be due to both the completion of myelination and an increase in the excitatory effectiveness of a given stimulus, made possible by increasing synaptic proliferation and neurotransmitter synthesis.

Synaptogenesis in the auditory cortex does not appear to have been studied in detail in any species. Data from the visual cortex may be comparable to the auditory cortex because in some species (e.g. cats) eye opening and middle ear pneumatization occur at the same time. Although, on electrophysiological criteria, the visual system of cats is somewhat less mature than the auditory system at birth and attains maturity at a later date (Ellingson and Wilcott, 1960), the sequences of synaptogenetic events in the two systems are likely to be rather similar. In the visual cortex there is postnatally a rapid decrease in neurone density and a rapid increase in neuronal volume; these take place before eye opening. The formation of synaptic contacts begins before eye opening but is most pronounced in the 2–3 weeks after the eyes have opened in cats and rabbits (Cragg, 1975; Vrensen *et al.*, 1977). During the fourth and fifth postnatal weeks in cats the development of the visual cortex is very sensitive to environmental influences, particularly in relation to competition between inputs from each eye (Hubel and Wiesel, 1970). Unlike the visual cortex, the auditory cortex is far removed from the initial site of convergence of information from the two receptor organs so it would not be subjected directly to developmental competition related to spatial representation. Given the importance of the spectral composition of auditory stimuli in determining the firing of adult auditory cortical neurones (see Chapter 7), it is possible that temporal features of the acoustic environment may be important influences during synaptogenesis in the auditory cortex.

As synapses develop, the dendritic arborizations of cortical neurones grow. Pyramidal cells of layer V are among the first to mature, and send their apical dendrites to layer I. At 6 weeks postnatal in dogs, neurones have their full complement of spinous dendritic processes (Fox, 1968). Damage to the cochlea at this time can lead to increases in the dendritic fields of aspinous stellate neurones in layers III–IV of the auditory cortex of rabbits, and a reduction in spine numbers on adjacent pyramidal cells (McMullen and Glaser, 1988; McMullen *et al.*, 1988).

The reciprocal, topographic connections in cats between the AI and MG or IC (described in Chapter 5) are present in the first postnatal week (Feng and Brugge, 1983). In contrast, connections made by callosal fibres, and the locations of the cell bodies projecting these fibres, require at least 7 weeks postnatally to achieve their adult dispositions (Feng and Brugge, 1983). Recall that in adult cats, callosal axons terminate as bands in layers III, V and VI, the cell bodies of origin lie in the same layers, and terminals are found in patches across the AI. In neonatal cats terminals occur throughout most of the deeper parts of the grey matter but it is not until the end of the third week that the vertical columns or patches may be observed, and still longer for the laminar concentration of terminals to be apparent. Similarly, the cells of origin of callosal fibres occupy continuous bands throughout layers III and V–VI at birth and it is only after about 50 postnatal days that the cells aggregate into patches within those laminae. This process of maturation appears not to be dependent on normal auditory input since it is unaltered by neonatal cochlear ablation (Brugge *et al.*, 1983). At 4 days postnatal most of the major extrinsic connections of the kitten auditory cortex are present (Cornwell *et al.*, 1984). During this early postnatal period projections from auditory to visual cortical areas are also present, but these are transitory and do not appear significantly in adults (DeHay *et al.*, 1988; Innocenti *et al.*, 1988).

For both cats and humans, the auditory system is ready to receive sound and initiate action potentials at birth. It has been shown that impulses can be evoked in the auditory cortex by electrical stimulation of the auditory pathway prior to the final maturation of the hair cell–auditory nerve synapse (Hassmannova and Myslivecek, 1967; Marty and Thomas, 1963). Let us now examine how the physiological properties of neurones change postnatally.

MATURATION OF DISCHARGE CHARACTERISTICS

Both slow-evoked potentials and single-unit responses can be recorded from the auditory cortex of cats at birth (König *et al.*, 1972) with suitably high intensity tonal stimuli. The evoked potentials have a positive–negative configuration superficially, and reverse at deeper levels to become negative–positive – this is essentially the same pattern as in adults. Unit discharges arise in relation to the negative phase of the evoked potential, so their latencies are shortest deeper in the cortex. At this early stage it is likely that the wiring of the cortical column is still proceeding, so that unit responses have shortest latencies near the white matter.

Pujol (1972) has studied the development of discharge patterns at a number of relays along the auditory pathway (including the auditory cortex) of anaesthetized cats. At all levels, most response patterns are of the 'onset' type in the first postnatal week. This is likely due at this stage to the immaturity of synapses and rapid depletion of neurotransmitter stores. In the second week, at lower levels rhythmic responses develop that follow the onset response; these rhythmic responses are replaced by sustained firing in subsequent days. Also, at a later time, inhibitory pauses are seen in firing patterns at the IC and MG. One interpretation is that local inhibitory circuits take longer to mature than principal cell relays. Probably as a consequence of these inhibitory pauses, most adult auditory cortical units continue to fire only at the stimulus onset in anaesthetized animals (Brugge *et al.*, 1988).

When midbrain auditory neurones first begin to respond to tonal stimuli they require high intensities at their CFs; the CF threshold gradually decreases in cats – those for IC neurones reach adult levels at about 4–5 weeks postnatal (Moore and Irvine, 1979). Tuning curves of units in the central nucleus of the IC in neonates show restriction in their frequency ranges compared with those in the dorsal cortex (Aitkin and Moore, 1975); they are, however, broader in bandwidth than their adult counterparts and achieve adult-like sharpness at 4 weeks (Moore and Irvine, 1979). The CFs of sequentially recorded units increase systematically as a function of dorsoventral location in the central nucleus of the IC of kittens. Such a tonotopic organization is qualitatively very similar to that seen in the adult (Merzenich and Reid, 1974) and is discernible at 5–6 days postnatal age, virtually as soon as CFs may be defined for units in the IC of kittens (Aitkin and Moore, 1975; Aitkin and Reynolds, 1975). Binaural classes – EE and EI – are also distinguishable in the first week, and it has been argued that both binaural connectivity and tonotopic organization are initially determined by innate, rather than environmental, influences (Aitkin and Reynolds, 1975).

The general sequence of developmental changes observed in the auditory cortex of kittens is similar to that described for the midbrain. In fact, according to Brugge *et al.* (1988), most of the age-related threshold changes can be accounted for by peripheral maturation. By the end of the second postnatal week CFs spanning the adult range (0.3–29 kHz) may be recorded, but thresholds continue to fall for a further 4 weeks. Tuning sharpness would be expected to increase during this period since tuning becomes sharper at lower levels; however, Brugge *et al.* (1988) illustrate some sharp tuning curves in their study in animals between 8 and 32 days of age.

Both monotonic and non-monotonic spike count functions of intensity can be recognized at least as early as 8 days postnatal, and adult-like sensitivity to interaural intensity differences may be demonstrated to occur at this time. The apparent relative immaturity of midbrain units at this age in terms of the upper limit of frequency responsiveness (Aitkin and Moore, 1975) and degree of sensitivity to interaural intensity differences (Moore and Irvine, 1981) is probably a result of sampling bias; the high-frequency representation at the cortical level is more accessible than the low-frequency representation in cats, and the reverse is the case for the inferior colliculus. Brugge *et al.* (1988) suggest that one should not view the development of the auditory pathway as only a serial phenomenon, and they propose a concept of functional circuit development (for example, the pathway processing interaural intensity differences) in which serial elements in fact develop at the same time.

Whatever the innate mechanism of central auditory processing may be, binaural properties, and maps of the representation of each ear in the auditory cortex, are subject to environmental events occurring in postnatal and adult life. A reduction in acoustic input created by reversibly blocking one ear canal shortly (1–4 days) after birth leads to relatively minor changes, other than an elevation of 10–20 dB in threshold, in the monaural influences of the operated ear, but to major changes in binaural interactions in the AI (Brugge *et al.*, 1985). In normal cats ipsilateral stimuli can effectively suppress the response to a contralateral stimulus when both are of similar intensity. In cats in which the ear ipsilateral to the AI is reversibly blocked for 8–14 months, the ipsilateral intensity needs to be as much as 40 dB greater than the contralateral intensity for it to be inhibitory (Brugge *et al.*, 1985). This effect may partly be due to the monaural threshold elevations, but may also reflect some degree of postnatal plasticity in binaural connectivity.

A more profound deficit in hearing is achieved by unilateral cochlear ablation in the neonatal period. In normal adult cats ipsilateral excitatory thresholds are some 10–15 dB higher than contralateral excitatory thresholds, and ipsilateral stimuli are excitatory for only about two thirds of neurones in the AI (Reale *et al.*, 1987). In adult cats with neonatal cochlear ablations, the thresholds of neurones ipsilateral to the non-operated ear are close to the normal contralateral values, and ipsilateral stimuli activate almost all neurones to provide a normal tonotopic map (Reale *et al.*, 1987). Similar changes have been observed in the inferior colliculus of gerbils following ablation of one cochlea early in postnatal life (Kitzes and Semple, 1985), so that the cortical changes may simply reflect those that have occurred at a lower level. Two of the hypotheses that might account for these changes have experimental support: (i) there are changes in the number of cells projecting from the cochlear nucleus to the ipsilateral inferior colliculus in gerbils with neonatal cochlear ablations, and (ii) ipsilateral connections may always be present but normally masked by contralateral input.

ADAPTABILITY OF THE MATURE AUDITORY CORTEX

We have seen that the auditory pathway of a neonate is organized basically like that of an adult, and that subsequent maturation is highly dependent on the acoustic environment to which the developing animal is exposed. To what extent is the final outcome of heredity and environment immutable in the adult animal? Can the organization of the auditory pathway be changed by alterations in the acoustic input to an adult? Introspection informs us that learning is not the prerogative of childhood but continues throughout adult life. If we correlate mental with cerebral events, as most neuroscientists do implicitly, then some form of neural plasticity associated with learning must occur in the brain throughout life. Furthermore, it is often considered that heightened sensitivity to sound occurs in adults deprived of vision, and there is anatomical evidence that the auditory cortex grows in mice deprived of vision (Gyllenstein *et al.*, 1966).

Weinberger and Diamond (1987) have reviewed the literature on single-unit plasticity in the auditory cortex and the physiological correlates of learning. Conditioning paradigms, imparting an auditory (conditioned) stimulus with extra-auditory significance, lead to changes in firing rate to that stimulus in two thirds of the units studied in the AI. Approximately equal numbers of units show increases and decreases in firing rates during conditioning. An even greater proportion of neurones in the AII show plasticity – only one of 22 units did not show changes during conditioning, and these changes in both regions were very rapid, taking between 5 and 17 trials to develop (Weinberger and Diamond, 1987). These quantitative changes (in firing rate) are likely to be generated by increased activity in the pre-existing pathways, since the time course of their development is too rapid to be accounted for by any structural plastic change. However, evidence is accumulating that major reorganizations may occur in certain cortical areas following changes to the peripheral receptor epithelium.

A series of studies by Merzenich, Kaas and co-workers (e.g. Kaas *et al.*, 1983) has shown that the maps of digit representation in the somatosensory cortex of primates can be altered in adults by the manipulation of cutaneous input from the hand. Cortical territory representing a certain digit can be occupied by input from neighbouring fingers when the digit is removed or its nerve supply sectioned. Overuse of a digit may lead to an expansion in its cortical representation. Somatosensory cortical maps appear highly plastic, at least in primates, although there is a limit to the expansion of a cortical territory following peripheral manipulations. It is not clear whether these changes are due to the disinhibition of normally present but normally masked synaptic inputs from neighbouring digits, to the increase in effectiveness or the proliferation of normally present but normally weak synapses, or to terminal sprouting from neighbouring zones. The latter possibility seems unlikely because the reorganization of the somatosensory cortex occurs virtually immediately after the peripheral manipulation, even though the maturation of sensitivity takes much longer.

Robertson and Irvine (1989) have examined the reorganization of the auditory cortex of adult guinea pigs following mechanical lesion of the cochlea on one side. They carried out both electrophysiological (compound auditory nerve action potential analyses) and histological (hair cell

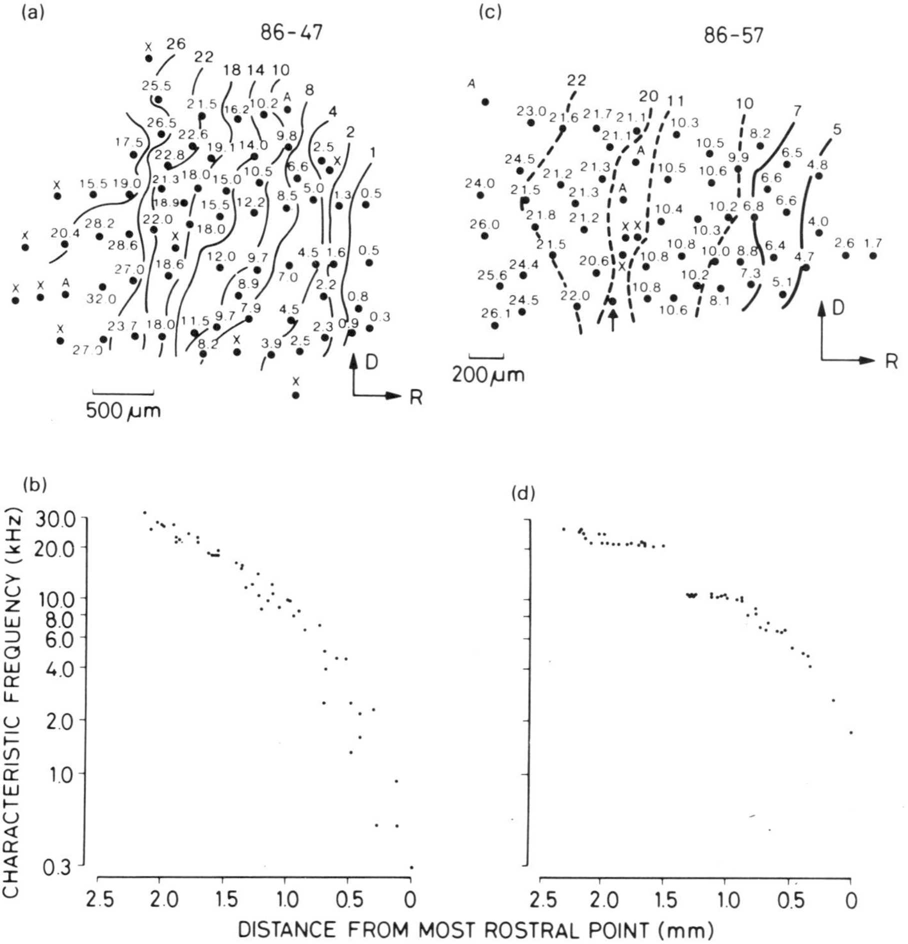

Figure 9.1 (a) Frequency organization of the auditory cortex in a normal guinea pig (86–47). Each dot represents the location on the cortical surface of an electrode penetration. Number beside each dot is the CF in kHz of clusters at that point. Points marked X were unresponsive to acoustic stimulation. Points marked A were acoustically responsive, but a clear CF could not be assigned. Lines are approximate isofrequency contours drawn by eye through loci of same CF as indicated by numeral at end of each line. R, rostral; D, dorsal. (b) Data in (a) plotted graphically; CF is plotted as a function of distance from the most rostral point at which a CF was assigned, for the rostral field only. Distance was measured along an axis as near as possible at right angles to the isofrequency contours. (c) Cortical frequency map. (d) Plot of CF against distance, for animal 86–57 which had received a cochlear lesion 35 days prior to mapping. Conventions as for map (a) except that broken lines in (c) delineate areas of expanded frequency representations with sensitive drive. Point marked with arrow in (c) had equally sensitive responses at 10.6 and 21.4 kHz. From Robertson and Irvine (1989).

counts) confirmations of the extents of their cochlear lesions, and mapped the auditory cortex of guinea pigs 35 to 81 days after the lesion. In another group of guinea pigs, partial tonotopic maps were made before and immediately after the lesion.

Recall (Chapter 4) that the auditory cortex of a guinea pig has two well-defined fields, one caudal and one rostral. Homologies with other animals suggest that the caudal field is the AI, but the rostral field has a more regular organization and was the main target of the study of Robertson and Irvine. The map of CFs in one normal animal, together with interpolated isofrequency contours is shown in Figure 9.1a. These data are replotted in Figure 9.1b as distance from the rostral end of the rostral field against CF, revealing an accelerating but monotonic increase in cortical distance occupied by progressively higher octaves, a finding common to most non-primate mammals.

A case with a cochlear lesion, causing a threshold loss of 50 dB between 11 and 20 kHz, is shown in Figure 9.1c,d. The rostral field has a similar rostro-caudal width to normal (2–2.5 mm) but lacks neurones with CFs between 11 and 20 kHz. A 200 μm wide strip between units of CF 10.3–10.8 and 20.6–21.3 is occupied by unresponsive or weakly responsive units, whereas the representations of CFs around 10 and 21 kHz are greatly expanded (Figure 9.1d). In this cortex much of the territory that would have represented 10–20 kHz has been occupied by input from segments of the organ of Corti flanking the lesion site.

Robertson and Irvine demonstrated that input from these flanking zones was already present when they remapped part of the auditory cortex immediately after a cochlear lesion. The sector that originally responded to the lesion frequency responded, at high threshold, to adjacent frequencies after the lesion was made. This suggests that the reorganization of the auditory cortex of guinea pigs is unlikely to be the result of terminal sprouting, confirming the findings for the somatosensory cortex.

This study is important in showing that the auditory cortex of adult guinea pigs is capable of modification as a result of environmental influences. Such modifications pose questions for applied research into the causes of human deafness and their remedies. If a human becomes profoundly deaf as an adult, what happens to the central auditory pathways of such a person? Do connections remain latent, awaiting activation, or is some of the auditory territory occupied by other neural systems? In the next chapter we shall examine more closely the human auditory cortex and the consequences of pathological changes to this region – cortical deafness.

10 *Cortical deafness*

In preceding chapters we have examined the anatomical structure, connectivity patterns and physiological properties of neurones in a cortical region defined as auditory initially on the basis of the effects of damage to this region. We should now look in detail at the evidence that this area is indeed necessary for normal hearing, and examine the behavioural and perceptual deficits following damage to the auditory cortex in animals and humans.

There has been a long tradition of behavioural experimentation on animals with lesions in parts of their brain auditory systems. The cortex, being relatively accessible, has been the subject of most of these studies, but there are problems in comparing the findings of lesion experiments. One relates to the locus and extent of a lesion. Until relatively recently, lesions have been placed in cats' auditory cortical areas defined on the basis of surface landmarks. For other species even less information has been available about the location of the auditory cortex and often none at all about subdivisions of this area. For these species 'temporal cortex' is more appropriate than 'auditory cortex', and this remains a problem for analysis of the effects of damage to the human cerebrum on hearing capacities.

Another problem in animal experimentation lies in the design of the experiment. We have discussed this in relation to sound localization experiments, some of which have used a left versus right discrimination and others a choice of one sound source location from many. In the case of frequency discrimination, there is less information available to the auditory system when two frequencies are separated by a period of silence than when the frequency of a continuous tone is altered. The behavioural response of the animal can also vary, from the animal going to a food reward, to a lever press or some form of conditioned suppression.

Finally, there are often uncertainties about whether or not a given loss is relearnable. If it is, then the target lesioned cannot be said to be necessary for the discrimination, even though it may normally be involved in processing for that discrimination. If it is not, it may be the loss of the relay presynaptic to the lesion (the MG in the case of the auditory cortex) that is important, not the auditory cortex itself.

Because of these variables, we will select those animal experiments in which complete degeneration of the ventral (or principal) division of the MG has been demonstrated, either unilaterally or bilaterally. This will only have occurred if all of the auditory cortical target areas of MG relay neurones have been removed; these, by definition, must include the AI. For human studies this condition can rarely be satisfied, partly because of the vagaries of the cerebrovascular insult or tumour, and partly because the state of the brain is unavailable for comment.

When is an animal considered to be 'deaf'? One criterion is that the animal should ignore a sound to which it has been conditioned to respond, or which is of biological significance to it. This is a necessary, but not sufficient, condition since this will also happen if the animal is asleep or paralysed. Thus, two further criteria are that (i) the animal is alert and, (ii) it is capable of normal movements, and that the appropriate coordinated movement required in the paradigm can be evoked by other stimuli. In the case of humans the task is much easier since the human can use some other means of communication to report deafness; nonetheless, the three criteria above must be satisfied.

Much of the behavioural work was carried out by Neff and co-workers in the 1950s and 1960s with cats. More recent work has included other species, different behavioural paradigms, and combinations of behavioural work with mapping studies. The basic paradigm employed in many of the studies by Neff's group (aside from those concerned with localization in space) is that of avoidance conditioning. The cat is trained to move from one side of a double-grill box to the other, when a warning sound is presented, to avoid a shock. A neutral signal, given alternately with the warning signal, indicates no shock, so it is possible to begin with neutral and warning signals that are considerably apart in frequency, intensity or pattern and, when the animal has learned a large difference, gradually to reduce the difference to a threshold discrimination. Once the animal is trained, the auditory cortex can be surgically removed and the animal retrained postoperatively.

UNILATERAL LESIONS OF THE AUDITORY CORTEX

A major deficit in sound localization ability in cats, consequent upon unilateral lesions of the AI, has been discussed in relation to the physiology of localization in Chapter 8. The revelation of this deficit is very dependent on the methods of testing used. Thus, when the animals are given a left–right choice, only small losses can be observed in cats with very large unilateral cortical lesions (Strominger, 1969), and dichotic listening tasks do not reveal substantial deficits in human listeners with unilateral temporal lobe tumours (Walsh, 1957).

The precise nature of a localization deficit may vary from one species to another. Kelly (1980) showed that rats can discriminate between left and right locations even in the absence of the entire auditory cortical fields of both sides. However, normal rats only appear capable of making fine localization distinctions for sounds located in the midline, and are virtually incapable of localizing laterally located sounds (Kavanagh and Kelly, 1986). In contrast, ferrets (carnivores) (Kavanagh and Kelly, 1987) and squirrel monkeys (primates) (Thompson and Cortez, 1983), like cats, make discrimination errors mainly in contralateral space when the auditory cortex is damaged on one side only.

Unilateral lesions of the auditory cortex do not appear to have any significant influence on pure tone frequency or intensity discrimination (e.g. Bocca *et al.*, 1955; Jerger *et al.*, 1969; Oesterreich *et al.*, 1971). Given the multiple representations of the cochlea in each hemisphere, such a finding is not surprising. However, some changes are observed in the discrimination of events that vary in time. Kaas *et al.* (1967), using the double-grill box technique, trained cats to cross the partition when the frequency of a tone changed from low to high (L to H). A special helmet enabled dichotic stimulation and the patterns of L (neutral) and H (warning) to each ear were varied, with the two ears never receiving tones simultaneously. When the auditory cortex was ablated unilaterally, the ability of the cat to make the discrimination depended on the pattern of binaural stimulation. If the cat was trained to make the discrimination to patterns presented to one ear, and the cortex contralateral to the ear was ablated, the habit was lost (but could be re-established with further training). The discrimination was retained if the cortex was removed ipsilateral to the 'trained ear'. A number of interpretations of this experiment are possible, one of which is that the analysis of binaural time patterns, as well as location in space, requires the integrity of both auditory cortices.

More recently, a direct demonstration has been made of a unilateral lesion deficit lateralized to the left temporal lobe. Macaques can be trained to discriminate between various macaque vocalizations, but this ability is lost transiently if the left primary and secondary auditory cortices are ablated, leading to severe or total degeneration of the MG (Heffner and Heffner, 1986b). Ablation of the right auditory cortex has no effect. Relearning of the discrimination after left temporal lobe lesions occurs after 5–15 sessions, but no relearning occurs if both left and right auditory cortices are ablated.

The perceptual effects of unilateral lesions of the auditory cortex in humans, 'hemianacousia' (Michel and Peronnet, 1980), are not commonly described in the literature, presumably because, in many cases, the patients may not be aware of a deficit, and only seek medical advice if the lesion affects the other hemisphere as well. The little that has been reported suggests that, as in the study of Kaas *et al.* with cats (1967), deficits in the interpretation of temporal information in the ear contralateral to the lesion may occur. Bocca *et al.* (1955) reported that temporal lobe tumours of one hemisphere do not affect tone or speech audiometry but lead to a poor discrimination of distorted voices in the contralateral ear. These deficits arose in relation to either left or right temporal lobe lesions, but aside from their unilateral nature it is probable that the tumours may have damaged highly variable amounts of auditory and periauditory cortex, including area 22. Nonetheless, observations by several other groups concur with the idea that perception of audioverbal material is impaired when presented to the ear contralateral to a unilateral auditory cortical lesion (Hecaen and Albert, 1978; Jerger *et al.*, 1969; Kimura, 1961). The likelihood of defining unilateral damage has been greatly increased by the introduction of computer scanning tomography and the recording of averaged evoked potentials from each hemisphere.

BILATERAL LESIONS OF THE AUDITORY CORTEX

Neff (1961) published a summary of the early work, much from his group, in which bilateral ablations of most or all of the auditory cortex had been carried out in cats. The question of whether the auditory cortex is necessary for frequency discrimination was the subject of a number of studies in the 1950s and early 1960s, and illustrates clearly the importance of experimental design, and the documentation of the lesion.

Meyer and Woolsey (1952) trained cats to discriminate between 3–10, 2 s long, 1000 Hz neutral tones, each separated by 1 s, and a 2 s long, 1100 Hz tone (warning). After the discrimination was learned, large bilateral removals of cerebral cortex were made. They found that even massive removals extending from the auditory cortex right back to the occipital lobe had no permanent effects and the discrimination could be relearned. However, if the second somatic area (SII) was included, there was a ‘catastrophic impairment’. Butler *et al.* (1957) examined discrimination of a similar frequency step, but used a continuous train of 1 s neutral tones interspersed with warning tones. In their experiment no effects on frequency discrimination occurred, even if the SII was removed bilaterally. Thompson (1960) attempted to resolve these differences by using the training procedures of both Meyer and Woolsey (1952) and Butler *et al.* (1957); Woolsey performed the operations. Thompson found that the same animals could, or could not, discriminate frequency depending on the conditions of training. Goldberg and Neff (1961) questioned both the training procedures and the extents of the lesions in the two earlier studies. Their lesions resulted in the retrograde degeneration of the entire MG and most of PO; their training schedule involved continuous pulses of 800 Hz (1 s on, 1 s off) alternately from two oscillators (neutral) until the second oscillator changed to 1000 Hz. The length of time the neutral signal was given was varied. Goldberg and Neff found that the frequency discrimination could be relearned if the neutral signal duration was 1 min or longer, but it could not be relearned with shorter duration neutral signals or if there was a silent interval between neutral and warning signals.

If all of the above studies are considered, one interpretation is that the discrimination between two different frequencies against a silent background (absolute frequency discrimination) cannot be relearned after bilateral auditory cortex ablation, but a change in the frequency of a continuous stimulus can be relearned. However, even the former hypothesis is questioned by Cranford *et al.* (1979), who have shown that bilaterally ablated cats can discriminate two tones of different frequency.

There has been less disagreement about the effects on thresholds and intensity discriminations of auditory cortex lesions. Initial effects are usually an elevation of threshold or difference limen, sometimes small (Cranford and Igarashi, 1977; Oesterreich *et al.*, 1971) (cats), sometimes large (Heffner and Heffner, 1986b) (macaque monkeys). If the hearing loss is followed over several months after the operation, recovery is always observed, although not necessarily complete recovery (Heffner and Heffner, 1986b; Kavanagh and Kelly, 1988; Ravizza and Masterton, 1972).

Kavanagh and Kelly (1988) used a reward conditioning paradigm to determine pure tone thresholds before and after bilateral removal of the primary auditory cortex of ferrets. The resultant hearing loss was variable but in all cases a substantial recovery occurred over several months; in two animals the recovery was virtually complete. In spite of this, these ferrets continued to be unable to localize sounds in lateral fields, so that such localization deficits cannot be explained in terms of general hearing loss.

In the case of macaque monkeys, whereas unilateral lesions have no obvious effect on hearing, bilateral lesions cause both a loss of discrimination of vocalizations and an elevation of threshold, in some cases of at least 70 dB, 3–4 weeks postoperatively (Heffner and Heffner, 1986a). Some improvement is seen at later times, but even after a year hearing losses of as much as 40 dB are still manifested by some monkeys.

The loss of discrimination of natural calls by macaque monkeys following bilateral ablations has already been mentioned. Somewhat similar, but less specific, effects are seen in cats when the temporal pattern of a sound is varied or when the animals are required to discriminate sounds of different durations (Neff, 1961). Diamond and Neff (1957) showed that a change in frequency from low, high, low, to high, low, high was not detected by cats with bilateral auditory cortex ablations. A deficit also occurs with a simpler low, high, low versus low, low, low discrimination (Diamond *et al.*, 1962). The ability of cats to discriminate between rising and falling frequency-modulated tones is impaired but never completely abolished by bilateral ablations (Kelly and Whitfield, 1971).

There are thus differences in the behavioural responses of mammals to bilateral ablation of the auditory cortex. Some caution should be exercised in generalizing about phylogenetic differences, but if the lesions in the various mammals have a similar extent, then it would seem that the primates are more dependent upon the integrity of the auditory cortex than are carnivores, and carnivores more than rats or opossums. Let us now extend these observations to humans.

BILATERAL TEMPORAL LOBE LESIONS IN HUMANS

One feature that stands out in case studies of humans with bilateral lesions is that no two cases are exactly the same. This is not surprising in view of the likely differences in the extent of lesions in each case. Partly as a consequence there is no complete agreement on the classification of particular disorders. It is interesting that, with behavioural experiments on animals, the experimenter can only comment with certainty on the aspect of hearing being examined by the particular conditioning paradigm in use. The experimenter has little knowledge about disorders of perception not being specifically investigated. In contrast the batteries of tests available to the neurologist and neuropsychologist can lead to an enormous documentation of a human case of deafness, and this can be augmented by computerized axial tomography (CAT), brain evoked response recording (BER), and of course routine audiometry. Two important features that make the animal experiment of particular use are the possibility of making highly localized and specifiable lesions, and of reconstructing post mortem the extent of the lesion in terms of local and remote degeneration. Even so, one cannot escape the thought that animals classified together as having similar lesions and similar behavioural response deficits may well be having very diverse perceptual experiences (if such occur in non-human animals) because of subtle differences between grossly similar lesions.

Michel and Peronnet (1980) make a simple distinction between the human phenomena of *cortical deafness*, when 'the patient feels deaf and seems to be so', and *auditory agnosia*, when 'the patient insists on the fact that he is not deaf'. Michel and Peronnet recognize the over-simplification of this distinction because many patients in both classes can still respond positively to audiometric testing. Most authors recognize two types of auditory agnosia. One is *pure word deafness*, in which semantic errors are made to the point where patients are unable to discriminate or comprehend spoken language, yet comprehension of non-verbal sounds is relatively normal. For a patient with pure word deafness, normal speech appears foreign, or so muffled that it cannot be heard clearly. In *non-verbal auditory agnosia*, however, speech recognition may be relatively normal but the nature of non-verbal sounds cannot be determined. These agnosias, in their purest form, are very rare. More commonly some aspects of each disorder occur following lesions of the temporal lobe.

A number of authors have hypothesized that pure word deafness may result from the disconnection of Wernicke's area from areas 41 and 42 (e.g. Bauer and Rubens, 1985). Thus a unilateral lesion of the left hemisphere may be as damaging for word recognition as a bilateral lesion (Gazzaniga *et al.*, 1973). However, destruction of that part of Heschl's gyrus containing the primary auditory cortex need not abolish non-verbal sound recognition, even if it does cause pure word deafness. Coslett *et al.* (1984) cite a case with bilateral primary auditory cortex infarcts, who had virtually zero verbal comprehension but had no trouble understanding and identifying non-verbal sounds. These authors suggest that this constitutes evidence for an input to the auditory association areas parallel to that ascending from the MG to the AI. On this basis, non-verbal sound comprehension would not be the product of hierarchical processing at the level of the auditory cortex.

Pure word deafness may exhibit as a part of a sequence of changes leading to, or recovering from, cortical deafness. For example, Mendez and Geehan (1988) studied two patients, one of whom recovered from cortical deafness to auditory agnosia, and the other who 'evolved' further to pure word deafness and finally to a mild disturbance of acoustic pattern recognition. The patient studied by Goldstein *et al.* (1975) initially showed thresholds of 70 dB at all frequencies. These improved over the next six months, as did her ability to identify non-verbal test sounds, but speech discrimination remained poor, so that her deficit had changed from cortical deafness to pure word deafness.

Graham *et al.* (1980) have reviewed the literature on cortical deafness in the light of individual case studies. Very few cases of total deafness of cerebral origin have been reported; most patients having bilateral temporal lobe lesions present with agnosic deficits. As noted above, these may change in severity with time and may sometimes largely recover.

The 12 cases reviewed by Graham *et al.* (1980) have diverse pathology but bilateral infarctions were common. Cortical deafness is usually diphasic, such that a transient deficit, related to unilateral damage, is followed by a sudden transient total deafness when the infarct spreads to both hemispheres (Jerger *et al.*, 1969). In such cases, pictures may be gained from CAT scans, in which the absorbance of X-rays is measured, indicating abnormal regions of low density that are attributable to haemorrhage. These pictures show damage in each temporal lobe that varies from one hemisphere to another and from one case to another, and is difficult to relate to cytoarchitectonic areas (e.g. Auberbach *et al.*, 1982; Earnest *et al.*, 1977; Graham *et al.*, 1980; Mendez and Geehan, 1988; Michel and Peronnet, 1980). In instances where cortical evoked response audiometry is employed, a lack of auditory-evoked cortical responses may correlate with lesions of the superior temporal lobe (Graham *et al.*, 1980; Michel and Peronnet, 1980). However, patients may be able to hear the test sounds yet have electrically unresponsive auditory cortices (Jerger *et al.*, 1969), or show evoked responses but be cortically 'deaf' (Earnest *et al.*, 1977).

Thus, in human studies, there is variation between observers in the classification of a deficit, in the extents of defects within a given classification, in the pathologies of the lesions and in the electrophysiological and audiometric analyses of different cases. One fact remains – that total deafness due to cortical damage is very rare and it may be transient.

An underlying feature of both animal studies and human lesion cases is that cortical damage interferes most with temporal pattern recognition. Humans with severely damaged temporal lobes may still hear pure tones and respond to an audiometric task. Similarly, cats can perform conditioned avoidance tasks linked to frequency and intensity discrimination, in the absence bilaterally of auditory cortex. Even in cases of total perceptual deafness, patients may show orienting responses to sound (Goldstein *et al.*, 1975) although Graham *et al.* (1980) question that this is a subcortical response. One speculative conclusion is the following: as more and more cerebral processing functions are shifted from subcortical to cortical levels as the phylogenetic tree is ascended, so the effects of cortical lesions become more dramatic for primates. It is probable that bilateral loss of corresponding auditory cortical areas will make humans more 'deaf' than cats.

THE CORTEX AND LANGUAGE PRODUCTION

It is not the intention of this book to delve deeply into the literature of psycho- or neuro-linguistics. A useful account of the anatomy of speech perception is given by Seldon (1985). It is, however, worth examining briefly the location and properties of cortex concerned with the production of vocalizations, and its relationship to the auditory cortex. The early history of the study of speech centres is of interest because it is also the early history of the localization of function in the cerebral cortex.

Broca (1888) examined the brains of two patients who, for various neurological reasons, had lost the ability to speak (aphasia) but appeared to have normal hearing. Broca concluded that aphasia occurred because of damage to a region subsequently shown to be anterior to the precentral gyral representation of the face, a region since referred to as Broca's motor speech area. Wernicke (1874), on the basis of a similarly small number of patients, located an auditory speech area just posterior to the auditory sensory area; I have referred to this previously in this book. A lesion in this region would also have effects on speech *production* because the speech produced by the patient would not be understood by him and he could not therefore correct it.

Since those times, our knowledge of the relations between the localization of lesions and specific aphasias has increased considerably, although it is very rare that lesions are restricted to areas as small as Broca's area. In general, the more the lesion encroaches on Broca's area, the greater the effect on speech production, while damage to Wernicke's area interferes with comprehension. Penfield and Roberts (1959), using the technique (described in detail below) of electrical stimulation of the brain of the conscious patient, showed that stimulation of the face representation in the precentral gyrus and in the supplementary motor area (superior and medial aspects of the precentral gyrus) could evoke vocalizations. However, no intelligible words were induced from an otherwise silent patient. Other stimuli, both within and outside the precentral motor area, and in the general vicinity of Broca's and Wernicke's areas, interfered with or arrested ongoing speech. There was little difference between the interfering effects of current when applied to Broca's, Wernicke's or the supplementary motor areas.

These studies, combined with the knowledge of the effects of lesions, provide the basis for the current ideas of the cortical events in vocalization. The intention to vocalize relates to activity in the supplementary motor area (the origin of the motor message to Broca's area) while feedback via the auditory cortex and the temporo-parietal (Wernicke's) area controls and corrects the signals passing from the motor centres to the muscles used in speech. These mechanisms normally reside in one hemisphere (usually the left) but the motor centre may shift to the opposite side following damage to the normal area.

At the time Penfield and Roberts (1959) published their work on human speech areas, the little that was known about vocalization control centres in other mammals suggested to Penfield and Roberts that 'other animals lack this inborn vocalization transmitting mechanism in the motor cortex' (p. 199). In recent years much information has been obtained about non-human primate communication from ethological, anatomical and physiological studies. There is evidence that vocalizations may not be used only for signalling simple emotional and behavioural contexts; conversations appear to take place between members of certain non-human primate species (e.g. Snowdon, 1988; Symmes and Biben, 1988). This would suggest a relatively elaborate, human-like cortical organization of vocalizations in these species.

In spite of this, the evidence indicates different cortical and subcortical pathways in humans and subhuman primates, and different roles in vocalization for areas in monkey brains corresponding anatomically to the motor speech, temporo-parietal speech and supplementary motor areas of humans (Ploog, 1988). Although an early study by Leyton and Sherrington (1917) suggested that brief sounds could be elicited in chimpanzees, electrical stimulation of the squirrel monkey's neocortex around the Sylvian fissure – including areas with anatomical positions corresponding to Broca's and Wernicke's areas – does not produce vocalization. Sound can be evoked, however, when the anterior limbic cortex of the squirrel monkey is stimulated.

Another major difference between humans and other primates is that lesions of the human cortical face area, involving Broca's area, lead to aphasia, whereas in monkeys a paralysis of tongue and lips occurs, but this does not prevent vocalization. This difference could be due to the fact that the direct connections to laryngeal motoneurones from pyramidal cells of the motor cortex of humans are absent in monkeys (Jürgens, 1976).

However, a similarity exists between the primate species in the role of the supplementary motor area. Damage to this region may produce prolonged aphasia in humans and cessation of certain spontaneous vocalizations in monkeys. It may play a role in the planning of vocalizations in all primate species.

THE RELATIONSHIP BETWEEN ELECTRICAL ACTIVITY AND HUMAN PERCEPTION

In the 1930s and 1940s, Penfield and co-workers introduced the technique of stimulation/mapping in an attempt to localize epileptogenic foci in the brain prior to their excision. Because their patients were conscious (locally anaesthetized) they were able to describe the sensations aroused by low-current stimuli applied to different points on the cortical surface. As mentioned in Chapter 1, Penfield and Rasmussen (1950) observed that stimulation of the anterior part of Heschl's gyrus gave rise to simple auditory sensations, confirming that this region was auditory cortex. As stated above, damage to this region can lead to deafness.

Rather different sensations occur when stimuli are applied to cortex posterior, anterior and lateral to the auditory area. Penfield and Perot (1963) refer to *experiential* responses, usually of voices, music or strings of words. When voices were heard, the majority were familiar, and these tended to be experienced more often with stimulation of the non-dominant temporal lobe. Sometimes a visual experience was associated with the voice – usually of the person who was speaking. Stimulation of the speech cortex adjacent to the auditory cortex produced an inhibition of speech (aphasia) but did not evoke vocalizations.

Experiential responses were usually qualitatively similar with repeated stimulations at the same point. Penfield and Perot referred to a case in another study in which a complex experiential response involving a number of voices interacting and communicating was evoked by electrical stimulation on 62 occasions. They concluded their study by contrasting this zone of interpretive cortex, from which experiential responses may be evoked, with the adjacent auditory cortex, stimulation of which evokes only simple sensations of buzzing or ringing. These fascinating studies of the mechanisms of human perception have long demanded some attempt at electrophysiological correlation. How do single units in the human temporal lobe respond to pure tones and speech sounds? Is there any feature-detection mechanism that could provide the basis for Penfield and co-workers' observations made using gentle electrical stimulation.

Creutzfeldt *et al.* (1987) have recorded the activity of single neurones in the lateral temporal lobe during listening and speaking. These recordings were made from areas due to be resected to alleviate severe seizures of temporal lobe origin. Patients were asked to listen to words drawn from a word list, to simple sounds, including a 1000 Hz pure tone, and to music, and were asked to read silently words that were displayed visually. Recordings were made from either hemisphere, and from at least two sites in 23 patients. These were distributed over the temporal gyrus but did not include Heschl's gyrus or the temporal speech area.

Although a variety of responses were seen, the following general conclusions were reached. First, the responses of all units were speech specific and were not evoked by simple stimuli, including the 1000 Hz tone used to alert the patient. Secondly, the semantic content of the words did not appear to be important – a unit could respond to a variety of words with different temporal and spectral compositions. Thirdly, attention was an important variable, but comprehension was not, and 'silent speech' did not produce any response. Finally, music appeared to suppress activity in the unit population examined.

Thus there is a correlation between electrical stimulation, which produces experiential sensations of vocalizations when applied to the temporal lobe, and unit responses to speech sounds but not simple sounds, when recordings are made in the same area. These responses, both experiential and electrophysiological, can be obtained from either temporal lobe. In this respect there is some discrepancy with studies of changes in regional cerebral blood flow during auditory discrimination. Roland *et al.* (1981) used the technique of isotope injection into the internal carotid artery combined with radioactivity measurements over one hemisphere, to measure the clearance of isotope and therefore the change in regional cerebral blood flow.

When subjects were asked to discriminate between contralateral tone pairs of varying temporal patterns, changes in blood flow compared with control conditions were observed, and these were greater in subjects whose right (non-dominant) hemisphere was examined compared with those in the left hemisphere in other subjects. However, if words rather than tone pairs were used, greater changes in blood flow occurred in the left (dominant) hemisphere (Nishizawa *et al.*, 1982). On this basis one would have predicted that units responding to words would be more commonly encountered in the left than in the right hemisphere. However, similar neuromagnetic responses to vowel onset occur in each hemisphere (Kaukoranta *et al.*, 1987) so perhaps the *volume* of tissue activated by speech sounds may be greater on the left than on the right hemisphere, rather than there being any qualitative differences in individual responses. It is clear that a strong direction of future auditory research will be towards elucidating the cortical mechanisms underlying vocalization discrimination in both animals and humans.

11 *Horizons*

Are we nearer to an understanding of the auditory cortex than we were 30 years ago? Certainly, for some aspects, the answer is strongly in the affirmative. We now know that, for a wide variety of species, including humans, a primary auditory cortex with precise cochleotopy is surrounded by secondary fields having varying degrees of precision in their wiring to the receptors of the inner ear. We know that, for some species, including echo-locating bats, there are functional subdivisions within these fields, important in specialized auditory perception. But we are yet to find compelling evidence, in species other than humans, for auditory cortex devoted to the analysis of species-specific vocalizations. This horizon is one of the most important beckoning to us, because this function of the auditory system – communication of abstract ideas through speech – is surely a major reason for the dominance of humans in the animal kingdom.

We have seen that the gross connectivity of the auditory cortex is established well before birth, with the timing of cell migration, axon growth and synaptogenesis varying from one species to another. The onset of hearing may occur *in utero* in some species; for others it occurs weeks after birth. The immediate postnatal auditory cortex is 'plastic', and a form of plasticity is still recognizable in the adult. A second horizon is the search for the determinants of such adult plasticity. For example, how is it that, many weeks after severe damage or lesions of the auditory cortex, thresholds improve and hearing may largely recover, even though the original neurones concerned with these functions have been irretrievably lost? The more we can discover about the reorganization of the cerebral cortex after injury or deprivation to a major system, the more we can guide research towards the alleviation of human suffering following cerebral vascular insults.

Are acoustic features encoded in the auditory cortex in such a way that an individual neurone has a specific and discrete function, so that its loss would lead to a specific and discrete deficit? The above observations on lesions would suggest not. Alternatively each acoustic feature may be represented by a *population* of neurones, the activity of which varies in time and space. Taking a perceptual leap forward, we can ask, 'What is the relationship between the neural activity evoked by sound and the perceptual phenomena we call "hearing"?' Do individual neurones, when active, cause individual perceptions to occur? Do perceptual phenomena arise in relation to populations of neurones? This horizon continues to be surveyed from at least two standpoints; those of the physiologist and of the philosopher (see Popper and Eccles, 1977).

Finally, in the future we can see that refinement and simplification will take place in techniques for studying the human auditory cortex. We have seen that PET techniques are an improvement over the original isotope wash-out methods because they enable auditory activity deep in the brain to be detected. Even more accuracy, at a cortical level, is provided by magnetoencephalography, and the cost of both this and the PET technique should diminish as technology advances. In the future, some of the questions about the auditory cortex will be answered by direct measurements on humans, but this will only have been made possible by the substantial basis to auditory neuroscience given by animal experimentation. One fact which is clear is that an understanding of the brain is emerging from the application of diverse experimental techniques to a variety of species, from the simplest forms to humans.

References

Abeles, M. and Goldstein, M. H. (1970) Functional architecture in cat primary auditory cortex: Columnar organization and organization according to depth. *Journal of Neurophysiology*, **33**, 172–187.

Abeles, M. and Goldstein, M. H. (1972) Responses of single units in the primary auditory cortex of the cat to tones and to tone pairs. *Brain Research*, **42**, 337–352.

Aitkin, L. M. (1976) Tonotopic organization at higher levels of the auditory pathway. In R. Porter (ed.) *International Review of Physiology. Neurophysiology II*, Vol. 10. University Park Press, Baltimore, pp. 249–279.

Aitkin, L. M. (1986) *The Auditory Midbrain*. Humana, Clifton, New Jersey, pp. 1–246.

Aitkin, L. M. (1990) Coding for auditory space. In M. J. Rowe and L. M. Aitkin (eds) *Information Processing in the Mammalian Auditory and Tactile Systems*. Alan R. Liss, New York, pp. 169–178.

Aitkin, L. M. and Dunlop, C. W. (1969) Inhibition in the medial geniculate body of the cat. *Experimental Brain Research*, **7**, 68–83.

Aitkin, L. M. and Gates, G. R. (1983) Connections of the auditory cortex of the brush-tailed possum, *Trichosurus vulpecula. Brain, Behavior and Evolution*, **22**, 75–88.

Aitkin, L. M. and Martin, R. L. (1987) The representation of stimulus azimuth by high best-frequency azimuth-selective neurons in the central nucleus of the inferior colliculus of the cat. *Journal of Neurophysiology*, **57**, 1185–1200.

Aitkin, L. M. and Moore, D. R. (1975) Inferior colliculus: II. Development of tuning characteristics and tonotopic organization in central nucleus of the neonatal cat. *Journal of Neurophysiology*, **38**, 1208–1216.

Aitkin, L. M. and Phillips, S. C. (1984) The interconnections of the inferior colliculi through their commissure. *Journal of Comparative Neurology*, **228**, 210–216.

Aitkin, L. M. and Prain, S. M. (1974) Medial geniculate body: Unit responses in the awake cat. *Journal of Neurophysiology*, **37**, 512–521.

Aitkin, L. M. and Reynolds, A. (1975) Development of binaural responses in the kitten inferior colliculus. *Neuroscience Letters*, **1**, 315–319.

Aitkin, L. M., Anderson, D. J. and Brugge, J. F. (1970) Tonotopic organization and discharge characteristics of single neurons in nuclei of the lateral lemniscus of the cat. *Journal of Neurophysiology*, **33**, 421–440.

Aitkin, L. M., Gates, G. R. and Phillips, S. C. (1984a) Responses of neurons in inferior colliculus to variations in sound-source azimuth. *Journal of Neurophysiology*, **52**, 1–17.

Aitkin, L. M., Irvine, D. R. F. and Webster, W. R. (1984b) Central neural mechanisms of hearing. In I. Darian-Smith (ed.) *Handbook of Physiology. Section 1: The Nervous System*. Volume III. *Sensory Processes*, Chap. 16. American Physiological Society, Bethesda, pp. 675–737.

Aitkin, L. M., Pettigrew, J. D., Calford, M. B., Phillips, S. C. and Wise, L. Z. (1985) Representation of stimulus azimuth by low-frequency neurons in inferior colliculus of the cat. *Journal of Neurophysiology*, **53**, 43–59.

Aitkin, L. M., Irvine, D. R. F., Nelson, J. E., Merzenich, M. M. and Clarey, J. C. (1986a) Frequency representation in the auditory midbrain and forebrain of a marsupial, the northern native cat (*Dasyurus hallucatis*). *Brain, Behavior and Evolution*, **29**, 17–28.

Aitkin, L. M., Merzenich, M. M., Irvine, D. R. F., Clarey, J. C. and Nelson, J. E. (1986b) Frequency representation in auditory cortex of the common marmoset (*Callithrix jacchus jacchus*). *Journal of Comparative Neurology*, **252**, 175–185.

Aitkin, L. M., Kudo, M. and Irvine, D. R. F. (1988a) Connections of the primary auditory cortex in the common marmoset, *Callithrix jacchus jacchus*. *Journal of Comparative Neurology*, **269**, 235–248.

Aitkin, L. M., Rajan, R., McKay, J. and Irvine, D. R. F. (1988b) Azimuthal sensitivities of units in the auditory cortex of cats. *Proceedings of the Australian Physiological and Pharmacological Society*, **19**, 112P.

Altman, J. and Bayer, S. A. (1981) Time of origin of neurons of the rat inferior colliculus and the relations between cytogenesis and tonotopic order in the auditory pathway. *Experimental Brain Research*, **42**, 411–423.

Altman, J. A., Syka, J. and Shmigidina, G. N. (1970) Neuronal activity in the medial geniculate body of the cat during monaural and binaural stimulation. *Experimental Brain Research*, **10**, 81–93.

Andersen, R. A., Knight, P. L. and Merzenich, M. M. (1980a) The thalamocortical and corticothalamic connections of AI, AII and the anterior auditory field (AAF) in the cat: Evidence for two largely segregated systems of connections. *Journal of Comparative Neurology*, **194**, 663–701.

Andersen, R. A., Snyder, R. L. and Merzenich, M. M. (1980b) The topographic organization of corticocollicular projections from physiologically identified loci in the AI, AII, and anterior auditory cortical fields of the cat. *Journal of Comparative Neurology*, **191**, 479–494.

Angevine, J. B. and Sidman, R. L. (1961) Autoradiographic study of cell migration during histogenesis of cerebral cortex in the mouse. *Nature*, **192**, 766–768.

Auberbach, S. H., Allard, T., Nasser, M., Alexander, M. P. and Albert, M. L. (1982) Pure word deafness: analysis of a case with bilateral lesions and a defect at the prephenomic level. *Brain*, **105**, 271–300.

Bauer, R. M. and Rubens, A. B. (1985) Agnosia. In K. M. Heilman and E. Valenstein (eds) *Clinical Neuropsychology*. Oxford University Press, New York, pp. 187–241.

Birnholz, J. C. and Benacerraf, B. R. (1983) The development of human fetal hearing. *Science*, **222**, 516–518.

Blasdel, G. G. and Salama, G. (1986) Voltage-sensitive dyes reveal a modular organization in monkey striate cortex. *Nature*, **321**, 579–585.

Bocca, E., Calearo, C., Cassindri, V. and Migliavocca, F. (1955) Testing 'cortical' hearing in temporal lobe tumours. *Acta Oto-laryngologica*, **45**, 289–304.

Boudreau, J. C. and Tsuchitani, C. (1968) Binaural interaction in the cat superior olive S segment. *Journal of Neurophysiology*, **31**, 442–454.

Bremer, F. (1958) Physiology of the corpus callosum. *Research Publications of the Association for Nervous and Mental Diseases*, **36**, 424–448.

Bremer, F. and Dow, R. S. (1939) The cerebral acoustic area of the cat, a combined oscillographic and cytoarchitectonic study. *Journal of Neurophysiology*, **2**, 308–318.

Broca, P. (1888) *Mémoires sur le Cerveau de l'Homme*. Reinwald, Paris, pp. 1–161.

Brodal, P. (1972) The corticopontine projection in the cat. The projection from the auditory cortex. *Archives Italiennes de Biologie*, **110**, 119–144.

Brodmann, K. (1909) *Vergleichende Lokalisationslehre der Grosshirnrinde*. Barth, Leipzig.

Brownell, W. E., Bader, C. R., Bertrand, D. and de Ribaupierre, Y. (1985) Evoked mechanical responses of isolated cochlear outer hair cells. *Science*, **227**, 194–196.

Brugge, J. F. (1982) Auditory cortical areas in primates. In C. N. Woolsey (ed.) *Cortical Sensory Organization. Vol. 3. Multiple Auditory Areas*. Humana, Clifton, New Jersey, pp. 59–70.

Brugge, J. F. and Merzenich, M. M. (1973) Responses of neurons in auditory cortex of the macaque monkey to monaural and binaural stimulation. *Journal of Neurophysiology*, **36**, 1138–1158.

Brugge, J. F., Dubrovsky, N. A., Aitkin, L. M. and Anderson, D. J. (1969) Sensitivity of single neurons in auditory cortex of cat to binaural tonal stimulation: effects of varying interaural time and intensity. *Journal of Neurophysiology*, **32**, 1005–1024.

Brugge, J. F., Feng, J. E., Reale, R. A. and Chan, J. C. (1983) Topographic distribution of primary auditory cortical field (AI) callosal axons in cats reared with unilateral or bilateral cochlear destruction. *Society for Neuroscience Abstracts*, **9**, 377.

Brugge, J. F., Orman, S. S., Coleman, J. R., Chan, J. C. K. and Phillips, D. P. (1985) Binaural interactions in cortical area AI of cats reared with unilateral atresia of the external ear canal. *Hearing Research*, **20**, 275–287.

Brugge, J. F., Reale, R. A. and Wilson, G. F. (1988) Sensitivity of auditory cortical neurons of kittens to monaural and binaural high frequency sound. *Hearing Research*, **34**, 127–140.

Butler, R. A., Diamond, I. T. and Neff, W. D. (1957) Role of auditory cortex in discrimination of changes of frequency. *Journal of Neurophysiology*, **20**, 108–120.

Caird, D. and Klinke, R. (1983) Processing of binaural stimuli by cat superior olivary complex neurons. *Experimental Brain Research*, **52**, 385–399.

Calford, M. B. and Webster, W. R. (1981) Auditory representation within principal division of cat medial geniculate body: An electrophysiological study. *Journal of Neurophysiology*, **45**, 1013–1028.

Calford, M. B., Webster, W. R. and Semple, M. N. (1983) Measurement of frequency selectivity of single neurons in the central auditory pathway. *Hearing Research*, **11**, 395–401.

Calford, M. B., Moore, D. R. and Hutchings, M. E. (1986) Central and peripheral contributions to the coding of acoustic space by neurons in the inferior colliculus of the cat. *Journal of Neurophysiology*, **55**, 587–603.

Campbell, A. W. (1905) *Histological Studies on the Localization of Cerebral Function*. Cambridge University Press, Cambridge.

Carlile, S. and Pettigrew, A. G. (1987) Directional properties of the auditory periphery in the guinea pig. *Hearing Research*, **31**, 111–122.

Celesia, G. G. (1976) Organization of auditory cortical areas in man. *Brain*, **99**, 403–414.

Chang, H.-T. (1950) The repetitive discharges of corticothalamic reverberating circuits. *Journal of Neurophysiology*, **13**, 235–257.

Chang, H.-T. (1953) Interaction of evoked cortical potentials. *Journal of Neurophysiology*, **16**, 133–144.

Code, R. A. and Winer, J. A. (1985) Commissural neurons in layer III of cat primary auditory cortex (AI): Pyramidal and non-pyramidal cell input. *Journal of Comparative Neurology*, **242**, 485–510.

Code, R. A. and Winer, J. A. (1986) Columnar organization and reciprocity of commissural connections in cat primary auditory cortex (AI). *Hearing Research*, **23**, 205–222.

Cook, C. J., Williams, C. and Gluckman, P. D. (1987) Brainstem auditory evoked potentials in the fetal sheep, in utero. *Journal of Developmental Physiology*, **9**, 429–439.

Cooper, M. L. and Rakic, P. (1981) Neurogenetic gradients in the superior and inferior colliculi of the Rhesus monkey. *Journal of Comparative Neurology*, **202**, 309–334.

Cornwell, P., Ravizza, R. and Payne, B. (1984) Extrinsic visual and auditory cortical connections in the 4-day-old kitten. *Journal of Comparative Neurology*, **229**, 97–120.

Coslett, H. B., Brashear, H. R. and Heilman, K. M. (1984) Pure word deafness after bilateral primary auditory cortex infarcts. *Neurology*, **34**, 347–352.

Cragg, B. G. (1975) The development of synapses in the visual system of the cat. *Journal of Comparative Neurology*, **160**, 147–166.

Cranford, J. L. and Igarashi, M. (1977) Effects of auditory cortex lesions on temporal summation in cats. *Brain Research*, **136**, 559–564.

Cranford, J. L., Igarashi, M. and Stramler, J. H. (1979) Effect of auditory neocortex ablation on pitch perception in the cat. *Journal of Neurophysiology*, **39**, 143–153.

Creutzfeldt, O. D., Hellweg, F.-C. and Schreiner, C. (1980) Thalamocortical transformation of responses to complex auditory stimuli. *Experimental Brain Research*, **39**, 87–104.

Creutzfeldt, O., Ojemann, G. and Lettich, E. (1987) Single neuron activity in the right and left human temporal lobe during listening and speaking. In J. Engel, G. A. Ojemann, H. O. Luders, and P. D. Williamson (eds) *Fundamental Mechanisms of Human Brain Function*. Raven Press, New York, pp. 69–81.

DeHay, C., Kennedy, H. and Bullier, J. (1988) Characterisation of transient cortical projections from auditory, somatosensory and motor cortices to visual areas 17, 18 and 19 in the kitten. *Journal of Comparative Neurology*, **272**, 68–89.

de Ribaupierre, F., Goldstein, M. H. and Yeni-Komshian, G. (1972) Intracellular study of the cat's primary auditory cortex. *Brain Research*, **48**, 185–204.

Diamond, I. T. and Neff, W. D. (1957) Ablation of temporal cortex and discrimination of auditory patterns. *Journal of Neurophysiology*, **20**, 300–315.

Diamond, I. T., Goldberg, J. M. and Neff, W. D. (1962) Tonal discrimination after ablation of auditory cortex. *Journal of Neurophysiology*, **25**, 223–235.

Diamond, I. T., Jones, E. G. and Powell, T. P. S. (1968a) Interhemispheric fiber connections of the auditory cortex of the cat. *Brain Research*, **11**, 177–193.

Diamond, I. T., Jones, E. G. and Powell, T. P. S. (1968b) The association connections of the auditory cortex of the cat. *Brain Research*, **11**, 560–579.

Diamond, I. T., Jones, E. G. and Powell, T. P. S. (1969) The projection of the auditory cortex upon the diencephalon and brain stem in the cat. *Brain Research*, **15**, 305–340.

Dickson, J. W. and Gerstein, G. L. (1974) Interaction between neurons in auditory cortex of the cat. *Journal of Neurophysiology*, **37**, 1239–1261.

Dowben, R. M. and Rose, J. E. (1953) A metal filled microelectrode. *Science*, **118**, 22.

Downman, C. B. B., Woolsey, C. N. and Lende, R. A. (1960) Auditory areas I, II and Ep: Cochlear representation, afferent paths and interconnections. *Bulletin of the Johns Hopkins Hospital*, **106**, 127–142.

Earnest, M. P., Monroe, P. A. and Yarnell, P. R. (1977) Cortical deafness – demonstration of the pathological anatomy by CT scan. *Neurology*, **27**, 1172–1175.

Ehret, G. (1983) Development of hearing and response behavior to sound stimuli: behavioral studies. In R. Romand (ed.) *Development of Auditory and Vestibular Systems*, Academic Press, New York, pp. 211–237.

Eisenman, L. M. (1974) Neural encoding of sound location: An electrophysiological study in auditory cortex (AI) of the cat using free field stimuli. *Brain Research*, **75**, 203–214.

Ellingson, R. J. and Willcott, R. C. (1960) Development of evoked responses in visual and auditory cortices of kittens. *Journal of Neurophysiology*, **23**, 363–375.

Erulkar, S. D., Rose, J. E. and Davies, P. W. (1956) Single unit activity in the auditory cortex of the cat. *Bulletin of the Johns Hopkins Hospital*, **99**, 55–86.

Espinosa, I. E. and Gerstein, G. L. (1988) Cortical auditory neuron interactions during presentation of 3-tone sequences: effective connectivity. *Brain Research*, **450**, 39–50.

Evans, E. F. (1981) The dynamic range problem, place and time coding at the level of cochlear nerve and nucleus. In J. Syka and L. M. Aitkin (eds) *Neuronal Mechanisms of Hearing*, Plenum, New York, pp. 69–85.

Evans, E. F. and Whitfield, I. C. (1964) Classification of unit responses in the auditory cortex of the unanaesthetized and unrestrained cat. *Journal of Physiology (London)*, **171**, 476–493.

Evans, E. F., Ross, H. F. and Whitfield, I. C. (1965) The spatial distribution of unit characteristic frequency in the primary auditory cortex of the cat. *Journal of Physiology (London)*, **179**, 238–247.

Feng, J. Z. and Brugge, J. F. (1983) Postnatal development of auditory callosal connections in the kitten. *Journal of Comparative Neurology*, **214**, 416–426.

Ferrier, D. (1876) *The Functions of the Brain*. Smith, Elder and Co., London.

FitzPatrick, K. A. and Imig, T. J. (1978) Projections of auditory cortex upon the thalamus and midbrain in the owl monkey. *Journal of Comparative Neurology*, **177**, 537–556.

FitzPatrick, K. A. and Imig, T. J. (1980) Auditory cortico-cortical connections in the owl monkey. *Journal of Comparative Neurology*, **192**, 589–610.

FitzPatrick, K. A. and Imig, T. J. (1982) Organization of auditory connections: The primate auditory cortex. In C. N. Woolsey (ed.) *Cortical Sensory Organization*, Vol. 3, Multiple Auditory Areas. Humana, Clifton, New Jersey, pp. 71–109.

Foote, S. L., Freedman, R. and Oliver, A. P. (1975) Effects of putative transmitters on neuronal activity in monkey auditory cortex. *Brain Research*, **86**, 229–242.

Forbes, B. F. and Moskowitz, N. M. (1974) Projections of auditory responsive cortex in the squirrel monkey. *Brain Research*, **67**, 239–254.

Fox, M. W. (1968) Neuronal development and ontogeny of evoked potentials in auditory and visual cortex of the dog. *Electroencephalography and Clinical Neurophysiology*, **24**, 213–226.

Funkenstein, H. H. and Winter, P. (1973) Responses to acoustic stimuli of units in the auditory cortex of awake squirrel monkeys. *Experimental Brain Research*, **18**, 464–488.

Galaburda, A. and Sanides, F. (1980) Cytoarchitectonic organization of the human auditory cortex. *Journal of Comparative Neurology*, **190**, 597–610.

Games, K. D. and Winer, J. A. (1988) Layer V in rat auditory cortex: projections to the inferior colliculus and contralateral cortex. *Hearing Research*, **34**, 1–25.

Gazzaniga, M. S. (1970) *The Bisected Brain*, Appleton, New York.

Gazzaniga, M. S., Glass, A. V., Sarno, M. T. and Posner, J. B. (1973) Pure word deafness and hemispheric dynamics – a case history. *Cortex*, **9**, 136–143.

Geisler, C. D., Rhode, W. S. and Hazelton, D. W. (1969) Responses of inferior colliculus neurons in the cat to binaural acoustic stimuli having wide-band spectra. *Journal of Neurophysiology*, **32**, 960–974.

Gerstein, G. L. and Kiang, N. Y.-S. (1964) Responses of single units in the auditory cortex. *Experimental Neurology*, **10**, 1–18.

Geschwind, N. and Levitsky, W. (1968) Human brain: left-right asymmetries in temporal speech region. *Science*, **161**, 186–187.

Gilbert, C. D. and Wiesel, T. N. (1981) Laminar specialization and intracortical connections in cat primary visual cortex. In F. O. Schmitt, F. G. Worden and F. Dennis (eds) *The Organization of the Cerebral Cortex*, M.I.T. Press, Cambridge, pp. 163–191.

Glaser, E. M., Van Der Loos, H. and Gissler, M. (1979) Tangential orientation and spatial order in dendrites of cat auditory cortex: a computer microscope study of Golgi impregnated material. *Experimental Brain Research*, **36**, 411–431.

Glass, I. and Wollberg, Z. (1979) Lability in the responses of cells in the auditory cortex of squirrel monkeys to species-specific vocalizations. *Experimental Brain Research*, **34**, 489–498.

Gold, T. (1948) Hearing II. The physical basis of the action of the cochlea. *Proceedings of the Royal Society, Series B*, **135**, 492–498.

Goldberg, J. M. and Brown, P. B. (1969) Response of binaural neurons of dog superior olivary complex to dichotic tonal stimuli: Some physiological mechanisms of sound localization. *Journal of Neurophysiology*, **32**, 613–636.

Goldberg, J. M. and Neff, W. D. (1961) Frequency discrimination after bilateral ablation of cortical auditory areas. *Journal of Neurophysiology*, **24**, 119–128.

Goldberg, J. M., Diamond, I. T. and Neff, W. D. (1958) Frequency discrimination after ablation of cortical projection areas of the auditory system. *Federation Proceedings*, **17**, 55.

Goldstein, M. H., Jr (1968) Single unit studies of cortical coding of simple acoustic stimuli. In F. D. Carlson (ed.) *Physiological and Biochemical*

Aspects of Nervous Integration. Prentice Hall, Englewood Cliffs, New Jersey, pp. 131–151.

Goldstein, M. H., Jr, Hall, J. L. and Butterfield, B. O. (1968) Single unit activity in the primary auditory cortex of unanesthetized cats. *Journal of the Acoustical Society of America*, **43**, 444–455.

Goldstein, M. H., Jr, Abeles, M., Daly, R. L. and McIntosh, J. (1970) Functional architecture in cat primary auditory cortex: Tonotopic organization. *Journal of Neurophysiology*, **33**, 188–197.

Goldstein, M. N., Brown, M. and Hollander, J. (1975) Auditory agnosia and cortical deafness: analysis of a case with a three-year followup. *Brain and Language*, **2**, 324–332.

Gonzalez-Lima, F. and Scheich, H. (1984) Classical conditioning enhances auditory 2-deoxyglucose patterns in the inferior colliculus. *Neuroscience Letters*, **51**, 79–84.

Graham, J., Greenwood, R. and Lelky, B. (1980) Cortical deafness. A case report and review of the literature. *Journal of Neurological Science*, **48**, 35–49.

Green, J. D. (1958) A simple microelectrode for recording from the central nervous system. *Nature*, **182**, 962.

Greenwood, D. D. (1961) Critical bandwidth and the frequency coordinates of the basilar membrane. *Journal of the Acoustical Society of America*, **33**, 1344–1356.

Gyllenstein, L., Malmfors, T. and Norrlin, M. L. (1966) Growth alteration in the auditory cortex of visually deprived mice. *Journal of Comparative Neurology*, **126**, 463–470.

Hall, J. L. and Goldstein, M. H. Jr (1968) Representation of binaural stimuli by single units in primary auditory cortex of unanesthetized cats. *Journal of the Acoustical Society of America*, **43**, 456–461.

Hand, P. J., Miselis, R. R. and Reivich, M. A. (1977) A ^{14}C-2-deoxyglucose metabolic mapping study of the rat posteromedial barrel subfield. *Society for Neuroscience Abstracts*, **3**, 483.

Hari, R. and Makela, J. P. (1988) Modification of neuromagnetic responses of the human auditory cortex by masking sounds. *Experimental Brain Research*, **71**, 87–92.

Hassmannova, J. and Myslivecek, J. (1967) Maturation of the primary cortical response to stimulation of medial geniculate body. *Electroencephalography and Clinical Neurophysiology*, **22**, 547–555.

Heath, C. J. and Jones, E. G. (1971) Interhemispheric pathways in the absence of a corpus callosum. *Journal of Anatomy*, **109**, 253–270.

Hecaen, H. and Albert, M. (1978) Disorders of auditory perception. In H. Hecaen and M. Albert (eds) *Human Neuropsychology*. John Wiley and Sons, New York, pp. 248–276.

Heffner, R. S. and Heffner, H. E. (1985) Hearing range of the domestic cat. *Hearing Research*, **19**, 85–88.

Heffner, H. E. and Heffner, R. S. (1986a) Hearing loss in Japanese macaques following bilateral auditory cortex lesions. *Journal of Neurophysiology*, **55**, 256–271.

Heffner, H. E. and Heffner, R. S. (1986b) Effect of unilateral and bilateral auditory cortex lesions on the discrimination of vocalizations by Japanese macaques. *Journal of Neurophysiology*, **56**, 683–701.

Hellweg, F. C., Koch, R. and Vollgrath, M. (1977) Representation of the cochlea in the neocortex of the guinea pig. *Experimental Brain Research*, **29**, 467–474.

Hind, J. E. (1953) An electrophysiological determination of tonotopic organization in the auditory cortex of the cat. *Journal of Neurophysiology*, **16**, 475–489.

Hind, J. E. (1960) Unit activity in the auditory cortex. In G. L. Rasmussen and W. F. Windle (eds) *Neural Mechanisms of the Auditory and Vestibular System*. Thomas, Springfield, pp. 201–210.

Horner, K. C., Servière, J. and Granier-Deferre, C. (1987) Deoxyglucose demonstration of in-utero hearing in the guinea pig foetus. *Hearing Research*, **26**, 327–333.

Hubel, D. H. (1957) Tungsten microelectrodes for recording single units. *Science*, **125**, 549–550.

Hubel, D. H. and Wiesel, T. N. (1962) Receptive fields, binocular interaction and functional architecture in the cat's visual cortex. *Journal of Physiology (London)*, **160**, 106–154.

Hubel, D. H. and Wiesel, T. N. (1963) Shape and arrangement of columns in cat's striate cortex. *Journal of Physiology (London)*, **165**, 559–568.

Hubel, D. H. and Wiesel, T. N. (1970) The period of susceptibility to the physiological effects of unilateral eye closure in kittens. *Journal of Physiology (London)*, **206**, 419–436.

Humphrey, A., Skeen, L. and Norton, T. (1980) Topographic organization of the orientation column system in the striate cortex of the tree shrew (Tupaia glis) II. Deoxyglucose mapping. *Journal of Comparative Neurology*, **192**, 549–566.

Hungerbühler, J. P., Saunders, J. C., Greenberg, J. and Reivich, M. (1981) Functional neuroanatomy of the auditory cortex studied with [2-^{14}C]-deoxyglucose. *Experimental Neurology*, **71**, 104–121.

Imig, T. J. and Adrian, H. O. (1977) Binaural columns in the primary field (AI) of cat auditory cortex. *Brain Research*, **138**, 241–257.

Imig, T. J. and Brugge, J. F. (1978) Sources and terminations of callosal axons related to binaural and frequency maps in primary auditory cortex of the cat. *Journal of Comparative Neurology*, **180**, 637–660.

Imig, T. J. and Morel, A. (1985) Tonotopic organization in ventral nucleus of medial geniculate body in the cat. *Journal of Neurophysiology*, **53**, 309–340.

Imig, T. J. and Reale, R. A. (1980) Patterns of cortico-cortical connections related to tonotopic maps in cat auditory cortex. *Journal of Comparative Neurology*, **192**, 293–332.

Imig, T. J. and Reale, R. A. (1981) Ipsilateral corticocortical projections related to binaural columns in cat primary auditory cortex. *Journal of Comparative Neurology*, **203**, 1–14.

Imig, T. J., Irons, W. A. and Samson, F. R. (1987) Azimuth and intensity selectivity of single neurons in field AI of cat auditory cortex. *Society for Neuroscience Abstracts*, **13**, 1468.

Imig, T. J., Ruggero, M. A., Kitzes, L. M., Javel, E. and Brugge, J. F. (1977) Organization of auditory cortex in the owl monkey (Aotus trivirgatus). *Journal of Comparative Neurology*, **171**, 111–128.

Innocenti, G. M., Berbel, P. and Clarke, S. (1988) Development of projections from auditory to visual areas in the cat. *Journal of Comparative Neurology*, **272**, 242–259.

Irvine, D. R. F. (1986) The auditory brainstem: A review of the structure and function of auditory brainstem processing mechanisms. In D. Ottoson (ed.) *Progress in Sensory Physiology*, Vol. 7. Springer-Verlag, Berlin, pp. 1–279.

Irvine, D. R. F. (1987) Interaural intensity differences in the cat: Changes in sound pressure level at the two ears associated with azimuthal displacements in the frontal horizontal plane. *Hearing Research*, **26**, 267–286.

Irvine, D. R. F. and Brugge, J. F. (1980) Afferent and efferent connections between the claustrum and parietal association cortex in cat: A horseradish peroxidase and autoradiographic study. *Neuroscience Letters*, **20**, 5–10.

Irvine, D. R. F. and Huebner, H. (1979) Acoustic response characteristics of neurons in non-specific areas of cat cerebral cortex. *Journal of Neurophysiology*, **42**, 107–122.

Irvine, D. R. F., Rajan, R. and Aitkin, L. M. (1989) A comparison of the sensitivity to azimuthal position and to interaural intensity differences of high-frequency neurones in auditory cortex of the cat. *Neuroscience Letters*, Suppl. 34, S99.

Ivarsson, C., de Ribaupierre, Y. and de Ribaupierre, F. (1988) Influence of auditory localization cues on neuronal activity in the auditory thalamus of the cat. *Journal of Neurophysiology*, **59**, 586–606.

Jackson, J. H. (1915) Hughlings Jackson on aphasia and kindred afflictions of speech. Selected writings and studies on speech and aphasia. *Brain*, **38**, 1–190.

Jenkins, W. M. and Masterton, R. B. (1982) Sound localization: Effects of unilateral lesions in central auditory system. *Journal of Neurophysiology*, **47**, 987–1016.

Jenkins, W. M. and Merzenich, M. M. (1984) Role of cat primary auditory cortex for sound-localization behavior. *Journal of Neurophysiology*, **52**, 819–847.

Jerger, J., Weikers, N. J., Sharbrough, F. W. and Jerger, S. (1969) Bilateral lesions of the temporal lobe. A case study. *Acta Oto-laryngologica*, Suppl. 258, 1–51.

Jones, E. G. and Powell, T. P. S. (1970) An anatomical study of converging sensory pathways within the cerebral cortex of the monkey. *Brain*, **93**, 793–820.

Jürgens, U. (1976) Projections from the cortical larynx area in the squirrel monkey. *Experimental Brain Research*, **25**, 401–411

Kaas, J., Axelrod, S. and Diamond, I. T. (1967) An ablation study of the auditory cortex in the cat using binaural tonal patterns. *Journal of Neurophysiology*, **30**, 710–724.

Kaas, J. H., Merzenich, M. M. and Killackey, H. P. (1983) The reorganization of somatosensory cortex following peripheral nerve damage in adult and developing mammals. *Annual Review of Neuroscience*, **6**, 325–356.

Karol, E. A. and Pandya, D. N. (1971) The distribution of the corpus callosum in the rhesus monkey. *Brain*, **94**, 471–486.

Katsuki, Y., Watanabe, T. and Maruyama, N. (1959) Activity of auditory neurons in upper levels of brain of cat. *Journal of Neurophysiology*, **22**, 343–359.

Kaufman, L. and Williamson, S. J. (1980) The evoked magnetic field of the human brain. *Annals of the New York Academy of Science*, **340**, 45–65.

Kaukoranta, E., Hari, R. and Lounasmaa, O. V. (1987) Responses of the human auditory cortex to vocal onset after fricative consonants. *Experimental Brain Research*, **69**, 19–23.

Kavanagh, G. L. and Kelly, J. B. (1986) Midline and lateral field sound localization in the albino rat (*Rattus norvegicus*). *Behavior Neuroscience*, **100**, 200–205.

Kavanagh, G. L. and Kelly, J. B. (1987) Contribution of auditory cortex to sound localization by the ferret (*Mustela putorius*). *Journal of Neurophysiology*, **57**, 1746–1766.

Kavanagh, G. L. and Kelly, J. B. (1988) Hearing in the ferret (*Mustela putorius*): effects of primary auditory cortical lesions on thresholds for pure tone detection. *Journal of Neurophysiology*, **60**, 879–888.

Kawamura, K. (1973) Corticocortical fiber connections of the cat cerebrum. I. The temporal region. *Brain Research*, **51**, 1–21.

Kawamura, K. (1975) The pontine projection from the inferior colliculus in the cat. An experimental anatomical study. *Brain Research*, **95**, 309–322.

Kelly, J. B. (1980) Effects of auditory cortical lesions on sound localization by the rat. *Journal of Neurophysiology*, **44**, 1161–1174.

Kelly, J. B. and Whitfield, I. C. (1971) Effects of auditory cortical lesions on discriminations of rising and falling frequency-modulated tones. *Journal of Neurophysiology*, **34**, 802–816.

Kelly, J. B., Judge, P. W. and Phillips, D. P. (1986) Representation of the cochlea in primary auditory cortex of the ferret (*Mustela putorius*). *Hearing Research*, **24**, 111–115.

Kelly, J. P. and Wong, D. (1981) Laminar connections of the cat's auditory cortex. *Brain Research*, **212**, 1–15.

Kiang, N. Y. S. (1984) Peripheral neural processing of auditory information. In I. Darian-Smith (ed.) *Handbook of Physiology*. Section 1. *The Nervous System*. Volume III. *Sensory Processes*, Chap. 15. American Physiological Society, Bethesda, pp. 639–674.

Kiang, N. Y. S., Rho, J. M., Northrop, C. C., Liberman, M. C. and Ryugo, D. K. (1982) Hair-cell innervation by spiral ganglion cells in adult cats. *Science*, **217**, 175–177.

Kimura, D. (1961) Some effects of temporal lobe damage on auditory perception. *Canadian Journal of Psychology*, **15**, 156–165.

King, A. J. and Palmer, A. R. (1983) Cells responsive to free-field auditory stimuli in guinea-pig superior colliculus: Distribution and response properties. *Journal of Physiology*, **342**, 361–381.

Kitzes, L. M. and Semple, M. N. (1985) Single unit responses in the inferior colliculus: Effects of neonatal cochlear ablation. *Journal of Neurophysiology*, **53**, 1483–1500.

Kitzes, L. M., Wrege, K. S. and Cassady, J. M. (1980) Patterns of responses of cortical cells to binaural stimulation. *Journal of Comparative Neurology*, **192**, 455–472.

Knight, P. L. (1977) Representation of the cochlea within the anterior auditory field (AAF) of the cat. *Brain Research*, **130**, 447–467.

Knudsen, E. I. and Konishi, M. (1978) Space and frequency are represented separately in auditory midbrain of the owl. *Journal of Neurophysiology*, **41**, 870–884.

König, N., Pujol, R. and Marty, R. (1972) A laminar study of evoked potentials and unit responses in the auditory cortex of the postnatal cat. *Brain Research*, **36**, 469–473.

Krmpotic-Nemanic, J., Kostovic, I., Kelovic, Z. and Nemanic, D. (1980) Development of acetylcholinesterase (AchE) staining in human fetal auditory cortex. *Acta Oto-laryngologica*, **89**, 388–392.

Krmpotic-Nemanic, J., Kostovic, I., Kelovic, Z., Nemanic, D. and Mrzljak, L. (1983) Development of the human fetal auditory cortex: growth of afferent fibres. *Acta Anatomy*, **116**, 69–73.

Kudo, M., Glendenning, K. K., Frost, S. B. and Masterton, R. B. (1986) Origin of mammalian thalamocortical projections. I. Telencephalic projections of the medial geniculate body in the opossum (*Didelphis virginiana*). *Journal of Comparative Neurology*, **245**, 176–197.

Kudo, M., Aitkin, L. M. and Nelson, J. E. (1989) Auditory forebrain organization of an Australian marsupial, the northern native cat (*Dasyurus hallucatus*). *Journal of Comparative Neurology*, **279**, 28–42.

Lauter, J. L., Herscovitch, P., Formby, C. and Raichle, M. E. (1985) Tonotopic organization in human auditory cortex revealed by positron emission tomography. *Hearing Research*, **20**, 199–205.

Lauter, J. L., Herscovitch, P. and Raichle, M. E. (1988) Human auditory physiology studied with positron emission tomography. In J. Syka and R. B. Masterton (eds) *Auditory Pathway*, Plenum, New York, pp. 313–317.

Leyton, A. S. F. and Sherrington, C. S. (1917) Observations on the excitable cortex of the chimpanzee, orang-utan and gorilla. *Quarterly Journal of Experimental Physiology*, **11**, 135–222.

Licklider, J. C. R. and Kryter, K. D. (1942) Frequency-localization in the auditory cortex of the monkey. *Federation Proceedings*, **1**, 51.

Ling, G. and Gerard, R. W. (1949) The normal membrane potential of frog sartorius fibres. *Journal of Cellular and Comparative Physiology*, **34**, 383–396.

Luethke, L. E., Krubitzer, L. A. and Kaas, J. H. (1988) Cortical connections of electrophysiologically and architectonically defined subdivisions of auditory cortex in squirrels. *Journal of Comparative Neurology*, **268**, 181–203.

McKenna, T., Ashe, J. H., Hui, G. K. and Weinberger, N. M. (1988) Muscarinic agonists modulate spontaneous and evoked unit discharge in auditory cortex of cat. *Synapse*, **2**, 54–68.

McMullen, N. T. and Glaser, E. M. (1982) Tonotopic organization of rabbit auditory cortex. *Experimental Neurology*, **75**, 208–220.

McMullen, N. T. and Glaser, E. M. (1988) Auditory cortical response to neonatal deafening: pyramidal neuron spine loss without changes in growth or orientation. *Experimental Brain Research*, **72**, 195–200.

McMullen, N. T., Goldberger, B., Suter, C. M. and Glaser, E. M. (1988) Neonatal deafening alters nonpyramidal dendrite orientation in auditory cortex: A computer microscope study in the rabbit. *Journal of Comparative Neurology*, **267**, 92–106.

Manley, J. A. and Müller-Preuss, P. (1978) Response variability of auditory cortex cells in the squirrel monkey to constant acoustic stimuli. *Experimental Brain Research*, **32**, 171–180.

Marin-Padilla, M. (1971) Early prenatal ontogenesis of the cerebral cortex (neocortex) of the cat (*Felis domestica*). A Golgi study. *Zeitschrift für Anatomie und Entwicklungsgeschichte*, **134**, 117–145.

Marty, R. and Thomas, J. (1963) Réponse électrocorticale à la stimulation du nerf cochléaire chez le chat nouveau-né. *Journale de Physiologie (Paris)*, **55**, 528–529.

Matsubara, J. A. and Phillips, D. P. (1988) Intracortical connections and their physiological correlates in the primary auditory cortex (AI) of the cat. *Journal of Comparative Neurology*, **268**, 38–48.

Mendelson, J. R. and Cynader, M. S. (1985) Sensitivity of cat primary auditory cortex (A1) neurons to the direction and rate of frequency modulation. *Brain Research*, **327**, 331–335.

Mendez, M. F. amd Geehan, G. R. (1988) Cortical auditory disorders – clinical and psychoacoustic features. *Journal of Neurology, Neurosurgery and Psychiatry*, **51**, 1–9.

Merzenich, M. M. and Brugge, J. F. (1973) Representation of the cochlear partition on the superior temporal plane of the macaque monkey. *Brain Research*, **50**, 275–296.

Merzenich, M. M. and Reid, M. D. (1974) Representation of the cochlea within the inferior colliculus of the cat. *Brain Research*, **77**, 397–415.

Merzenich, M. M., Knight, P. L. and Roth, G. L. (1975) Representation of cochlea within primary auditory cortex in the cat. *Journal of Neurophysiology*, **38**, 231–249.

Merzenich, M. M., Kaas, J. H. and Roth, G. L. (1976) Auditory cortex in the grey squirrel: Tonotopic organization and architectonic fields. *Journal of Comparative Neurology*, **166**, 387–401.

Merzenich, M. M., Colwell, S. A. and Andersen, R. A. (1982) Auditory forebrain organization. In C. N. Woolsey (ed.) *Cortical Sensory Organization*, Vol. 3. *Multiple Auditory Areas*, Humana, Clifton, New Jersey, pp. 43–57.

Mettler, F. A. (1935) Corticofugal fiber connections of the cortex of *Macaca mulatta*. d. The temporal region. *Journal of Comparative Neurology*, **63**, 25–47.

Meyer, D. R. and Woolsey, C. N. (1952) Effects of localized cortical destruction on auditory discriminative conditioning in cat. *Journal of Neurophysiology*, **15**, 149–162.

Michel, F. and Peronnet, F. (1980) A case of cortical deafness: clinical and electrophysiological data. *Brain and Language*, **10**, 367–377.

Middlebrooks, J. C. and Knudsen, E. I. (1984) A neural code for auditory space in the cat's superior colliculus. *Journal of Neuroscience*, **4**, 2621–2634.

Middlebrooks, J. C. and Pettigrew, J. D. (1981) Functional classes of neurons in primary auditory cortex of the cat distinguished by sensitivity to sound location. *Journal of Neuroscience*, **1**, 107–120.

Middlebrooks, J. C. and Zook, J. M. (1983) Intrinsic organization of the cat's medial geniculate body identified by projections to binaural response-specific bands in the primary auditory cortex. *Journal of Neuroscience*, **3**, 203–224.

Middlebrooks, J. C., Dykes, R. W. and Merzenich, M. M. (1980) Binaural response-specific bands in primary auditory cortex (AI) of the cat: Topographical organization orthogonal to isofrequency contours. *Brain Research*, **181**, 31–48.

Miller, J. M., Sutton, D., Pfingst, B., Ryan, A., Beaton, R. and Gourevitch, G. (1972) Single cell activity in the auditory cortex of rhesus monkeys: Behavioral dependency. *Science*, **177**, 449–451.

Mitani, A. and Shimokouchi, M. (1985) Neuronal connections in the primary auditory cortex: an electrophysiological study in the cat. *Journal of Comparative Neurology*, **235**, 417–429.

Mitani, A., Shimokouchi, M., Itoh, K., Nomura, S., Kudo, M. and Mizuno, N. (1985) Morphology and laminar organization of electrophysiologically identified neurons in the primary auditory cortex in the cat. *Journal of Comparative Neurology*, **235**, 430–447.

Mitani, A., Shimokouchi, M. and Nomura, S. (1983) Effects of stimulation of the primary auditory cortex upon colliculogeniculate neurons in the inferior colliculus of the cat. *Neuroscience Letters*, **42**, 185–189.
Moore, C. N., Casseday, J. H. and Neff, W. D. (1974) Sound localization: the role of the commissural pathways in the auditory system of the cat. *Brain Research*, **82**, 13–26.
Moore, D. R. (1982) Late onset of hearing in the ferret. *Brain Research*, **253**, 309–311.
Moore, D. R. and Irvine, D. R. F.(1979) The development of some peripheral and central auditory responses in the neonatal cat. *Brain Research*, **163**, 49–60.
Moore, D. R. and Irvine, D. R. F. (1981) Development of responses to acoustic interaural intensity differences in the cat inferior colliculus. *Experimental Brain Research*, **41**, 301–309.
Moore, R. Y. and Goldberg, J. M. (1963) Ascending projections of the inferior colliculus in the cat. *Journal of Comparative Neurology*, **121**, 109–136.
Morel, A. and Imig, T. J. (1987) Thalamic projections to fields A, AI, P, and VP in the cat auditory cortex. *Journal of Comparative Neurology*, **265**, 119–144.
Morest, D. K. (1964) The neuronal architecture of the medial geniculate body of the cat. *Journal of Anatomy, London*, **98**, 611–630.
Morest, D. K. (1965) The laminar structure of the medial geniculate body of the cat. *Journal of Anatomy, London*, **99**, 143–160.
Morest, D. K. (1971) Dendrodendritic synapses of cells that have axons: The fine structure of the Golgi type II cell in the medial geniculate body of the cat. *Zeitschrift für Anatomie und Entwicklungsgeschichte*, **133**, 216–246.
Morest, D. K. (1975) Synaptic relationship of Golgi type II cells in the medial geniculate body of the cat. *Journal of Comparative Neurology*, **162**, 157–194.
Morrell, F. (1972) Visual system's view of acoustic space. *Nature*, **238**, 44–46.
Mountcastle, V. B. (1957) Modality and topographic properties of single neurons of cat's somatic sensory cortex. *Journal of Neurophysiology*, **20**, 408–434.
Müller, C. M. and Scheich, H. (1988) Contribution of GABA-ergic inhibition to the response characteristics of auditory units in the avian forebrain. *Journal of Neurophysiology*, **59**, 1673–1689.
Neff, W. D. (1961) Neural mechanisms of auditory discrimination. In W. Rosenblith (ed.) *Sensory Communication*. Technology Press and Wiley, New York, pp. 259–278.
Neff, W. D. (1968) Localization and lateralization of sound in space. In A. V. S. de Reuck and J. Knight (eds) *Hearing Mechanisms in Vertebrates*. Churchill, London, pp. 207–233.
Neff, W. D. and Diamond, I. T. (1958) The neural basis of auditory discrimination. In H. F. Harlow and C. N. Woolsey (eds) *Biological and Biochemical Bases of Behavior*. University of Wisconsin Press, Madison, pp. 101–126.
Neff, W. D. and Hind, J. E. (1955) Auditory thresholds of the cat. *Journal of the Acoustical Society of America*, **27**, 480–483.
Newman, J. D. and Wollberg, Z. (1973) Multiple coding of species-specific vocalizations in the auditory cortex of squirrel monkeys. *Brain Research*, **54**, 287–304.
Nishizawa, Y., Olsen, T. S., Larsen, B. and Lassen, N. A. (1982) Left-right cortical asymmetries of regional cerebral blood flow during listening to

words. *Journal of Neurophysiology*, **48**, 458–466.

Nomoto, M. (1980) Discharge patterns of the primary auditory cortex in cats. *Japanese Journal of Physiology*, **30**, 427–442.

Oertel, D. (1985) Use of brain slices in the study of the auditory system: spatial and temporal summation of synaptic inputs in cells of the anteroventral cochlear nucleus of the mouse. *Journal of the Acoustical Society of America*, **78**, 328–333.

Oesterreich, R. E., Strominger, N. L. and Neff, W. D. (1971) Neural structures mediating differential sound intensity discrimination in the cat. *Brain Research*, **27**, 251–270.

Oliver, D. L. and Morest, D. K. (1984) The central nucleus of the inferior colliculus in the cat. *Journal of Comparative Neurology*, **222**, 237–264.

Oonishi, S. and Katsuki, Y. (1965) Functional organization and integrative mechanism of the auditory cortex of the cat. *Japanese Journal of Physiology*, **15**, 342–365.

Osen, K. K. and Mugnaini, E. (1981) Neuronal circuits in the dorsal cochlear nucleus. In J. Syka and L. M. Aitkin (eds) *Neuronal Mechanisms of Hearing*. Plenum, New York, pp. 119–125.

Pandya, D. N., Hallett, M. and Mukherjee, S. K. (1969) Intra- and interhemispheric connections of the neocortical auditory system in the rhesus monkey. *Brain Research*, **14**, 49–65.

Paula-Barbosa, M. M. and Sousa-Pinto, A. (1973) Auditory cortical projections to the superior colliculus in the cat. *Brain Research*, **50**, 47–61.

Pelleg-Toiba, R. and Wollberg, Z. (1989) Tuning properties of auditory cortex cells in awake squirrel monkeys. *Experimental Brain Research*, **74**, 353–364.

Penfield, W. (1958) *The Excitable Cortex in Conscious Man*. Liverpool University Press, Liverpool.

Penfield, W. and Perot, P. (1963) The brain's record of auditory and visual experiences. *Brain*, **86**, 595–696.

Penfield, W. and Rasmussen, T. (1950) *The Cerebral Cortex of Man: a Clinical Study of Localization of Function*. Macmillan, New York.

Penfield, W. and Roberts, L. (1959) *Speech and Brain Mechanisms*. Princeton University Press, Princeton, New Jersey.

Pfingst, B. E., O'Connor, T. A. and Miller, J. M. (1977) Response plasticity of neurons in auditory cortex of the rhesus monkey. *Experimental Brain Research*, **29**, 393–404.

Phillips, D. P. (1987) Stimulus intensity and loudness recruitment: Neural correlates. *Journal of the Acoustical Society of America*, **82**, 1–12.

Phillips, D. P. (1988) Effect of tone-pulse rise time on rate-level functions of cat auditory cortex neurons: excitatory and inhibitory processes shaping responses to tone onset. *Journal of Neurophysiology*, **59**, 1524–1539.

Phillips, D. P. and Cynader, M. S. (1985) Some neural mechanisms in the cat's auditory cortex underlying sensitivity to combined tone and wide-spectrum noise stimuli. *Hearing Research*, **18**, 87–102.

Phillips, D. P. and Gates, G. R. (1982) Representation of the two ears in the auditory cortex: A re-examination. *International Journal of Neuroscience*, **16**, 41–46.

Phillips, D. P. and Hall, S. E. (1987) Responses of single neurons in cat auditory cortex to time-varying stimuli: linear amplitude modulations. *Experimental Brain Research*, **67**, 479–492.

Phillips, D. P. and Irvine, D. R. F. (1979a) Acoustic input to single neurons in pulvinar–posterior complex of cat thalamus. *Journal of Neurophysiology*, **42**, 123–136.

Phillips, D. P. and Irvine, D. R. F. (1979b) Methodological considerations in mapping auditory cortex: binaural columns in AI of cat. *Brain Research*, **161**, 342–346.

Phillips, D. P. and Irvine, D. R. F. (1981) Response of single neurons in physiologically-defined area AI of cat cerebral cortex: Sensitivity to interaural intensity differences. *Hearing Research*, **4**, 299–308.

Phillips, D. P. and Irvine, D. R. F. (1982) Properties of single neurons in the anterior auditory field (AAF) of cat cerebral cortex. *Brain Research*, **248**, 237–244.

Phillips, D. P. and Irvine, D. R. F. (1983) Some features of binaural input to single neurons in physiologically defined area AI of cat cerebral cortex. *Journal of Neurophysiology*, **49**, 383–395.

Phillips, D. P. and Orman, S. S. (1984) Responses of single neurons in posterior field of cat auditory cortex to tonal stimulation. *Journal of Neurophysiology*, **51**, 147–163.

Phillips, D. P., Calford, M. B., Pettigrew, J. D., Aitkin, L. M. and Semple, M. N. (1982) Directionality of sound pressure transformation at the cat's pinna. *Hearing Research*, **8**, 13–28.

Phillips, D. P., Mendelson, J. R., Cynader, M. S. and Douglas, R. M. (1985a) Responses of single neurones in cat auditory cortex to time-varying stimuli: frequency-modulated tones of narrow excursion. *Experimental Brain Research*, **58**, 443–454.

Phillips, D. P., Orman, S. S., Musicant, A. D. and Wilson, G. F. (1985b) Neurons in the cat's primary auditory cortex distinguished by their responses to tones and wide-spectrum noise. *Hearing Research*, **18**, 73–86.

Pickles, J. O. (1982) *An Introduction to the Physiology of Hearing*. Academic Press, London.

Ploog, D. (1988) Neurobiology and pathology of subhuman vocal communication. In D. Todt, P. Goedeking and D. Symmes (eds) *Primate Vocal Communication*. Springer, Berlin, pp. 195–212.

Poljak, S. (1932) Origin, course, termination and internal organization of the auditory radiation. In H. M. Evans and I. M. Thompson (eds) *The Main Afferent Fiber Systems of the Cerebral Cortex in Primates*. University of California Press, Berkeley, pp. 81–104.

Popper, K. R. and Eccles, J. C. (1977) *The Self and Its Brain*. Springer International, Berlin.

Pujol, R. (1972) Development of tone-burst responses along the auditory pathway in the cat. *Acta Oto-laryngologica*, **74**, 383–391.

Pujol, R. and Hilding, D. (1973) Anatomy and physiology of the onset of auditory function. *Acta Oto-laryngologica*, **76**, 1–10.

Raichle, M. E. (1983) Positron emission tomography. *Annual Review of Neuroscience*, **6**, 249–267.

Ravizza, R. J. and Masterton, B. (1972) Contribution of neocortex to sound localization in opossum (*Didelphis virginiana*). *Journal of Neurophysiology*, **35**, 344–356.

Reale, R. A. and Imig, T. J. (1980) Tonotopic organization in auditory cortex of the cat. *Journal of Comparative Neurology*, **192**, 265–291.

Reale, R. A. and Imig, T. J. (1983) Auditory cortical field projections to the basal ganglia of the cat. *Neuroscience*, **8**, 67–86.

Reale, R. A. and Kettner, R. E. (1986) Topography of binaural organization in primary auditory cortex of the cat: Effects of changing interaural intensity. *Journal of Neurophysiology*, **56**, 663–682.

Reale, R. A., Brugge, J. F. and Feng, J. Z. (1983) Geometry and orientation of neuronal processes in cat primary auditory cortex (AI) related to characteristic-frequency maps. *Proceedings of the National Academy of Science USA*, **80**, 5449–5453.

Reale, R. A., Brugge, J. F. and Chan, J. C. K. (1987) Maps of auditory cortex in cats reared after unilateral cochlear ablation in the neonatal period. *Developmental Brain Research*, **34**, 281–290.

Robertson, D. and Irvine, D. R. F. (1989) Plasticity of frequency organization in auditory cortex of guinea pigs with partial unilateral deafness. *Journal of Comparative Neurology*, **282**, 456–471.

Roland, P. E., Skinhoj, E. and Lassen, N. A. (1981) Focal activations of human cerebral cortex during auditory discrimination. *Journal of Neurophysiology*, **45**, 1139–1151.

Romani, G. L., Williamson, S. J. and Kaufman, L. (1982) Tonotopic organization of the human auditory cortex. *Science*, **216**, 1339–1340.

Rose, J. E. (1949) The cellular structure of the auditory region of the cat. *Journal of Comparative Neurology*, **91**, 409–440.

Rose, J. E. and Woolsey, C. N. (1949) The relations of thalamic connections, cellular structure and evocable electrical activity in the auditory region of the cat. *Journal of Comparative Neurology*, **91**, 441–466.

Rose, J. E. and Woolsey, C. N. (1958) Cortical connections and functional organization of the thalamic auditory system of the cat. In H. F. Harlow and C. N. Woolsey (eds) *Biological and Biochemical Bases of Behavior*. University of Wisconsin Press, Madison, pp. 127–150.

Rose, J. E., Greenwood, D. D., Goldberg, J. M. and Hind, J. E. (1963) Some discharge characteristics of single neurons in the inferior colliculus of the cat. I. Tonotopical organization, relation of spike-counts to tone intensity, and firing patterns of single elements. *Journal of Neurophysiology*, **26**, 294–320.

Rose, J. E., Brugge, J. F., Anderson, D. J. and Hind, J. E. (1967) Phase-locked response to low-frequency tones in single auditory nerve fibers of the squirrel monkey. *Journal of Neurophysiology*, **30**, 769–793.

Roth, G. L., Aitkin, L. M., Andersen, R. A. and Merzenich, M. M. (1978) Some features of the spatial organization of the central nucleus of the inferior colliculus of the cat. *Journal of Comparative Neurology*, **182**, 661–680.

Roth, G. L., Kochhar, R. K. and Hind, J. E. (1980) Interaural time differences: Implications regarding the neurophysiology of sound localization. *Journal of the Acoustical Society of America*, **68**, 1643–1651.

Rouiller, E. M., Hornung, J. P. and de Ribaupierre, F. (1989) Extrathalamic ascending projections to physiologically identified fields of the cat auditory cortex. *Hearing Research*, **40**, 233–246.

Ryan, A. F., Woolf, N. K. and Sharp, F. B. (1982) Tonotopic organization in the central auditory pathway of the mongolian gerbil: a 2-deoxyglucose study. *Journal of Comparative Neurology*, **207**, 369–381.

Sally, S. L. and Kelly, J. B. (1988) Organization of auditory cortex in the albino rat: sound frequency. *Journal of Neurophysiology*, **59**, 1627–1638.

Sanchez-Longo, L. P. and Forster, F. M. (1958) Clinical significance of impairment of sound localization. *Neurology*, **8**, 119–125.

Sanides, F. (1975) Comparative neurology of the temporal lobe in primates, including man with reference to speech. *Brain and Language*, **2**, 396–419.

Scheich, H. (1983) Two columnar systems in the auditory neostriatum of the chick: Evidence from 2-deoxyglucose. *Experimental Brain Research*, **51**, 199–205.

Schreiner, C. E. and Cynader, M. S. (1984) Basic functional organization of second auditory cortical field (AII) of the cat. *Journal of Neurophysiology*, **51**, 1284–1305.

Seldon, H. L. (1981a) Structure of human auditory cortex. I. Cytoarchitectonics and dendritic distributions. *Brain Research*, **229**, 227–294.

Seldon, H. L. (1981b) Structure of human auditory cortex. II. Axon distributions and morphological correlates of speech perception. *Brain Research*, **229**, 295–310.

Seldon, H. L. (1985) The anatomy of speech perception: Human auditory cortex. In A. Peters and E. G. Jones (eds) *Cerebral Cortex*, Vol. 4. Plenum, New York, pp. 273–327.

Seltzer, B. and Pandya, D. N. (1978) Afferent cortical connections and architectonics of the superior temporal sulcus and surrounding cortex in the rhesus monkey. *Brain Research*, **149**, 1–24.

Semple, M. N. and Aitkin, L. M. (1979) Representation of sound frequency and laterality by units in central nucleus of cat inferior colliculus. *Journal of Neurophysiology*, **42**, 1626–1639.

Semple, M. N. and Kitzes, L. M. (1985) Single-unit responses in the inferior colliculus: Different consequences of contralateral and ipsilateral auditory stimulation. *Journal of Neurophysiology*, **53**, 1467–1482.

Semple, M. N. and Kitzes, L. M. (1987a) Binaural processing of sound pressure level in the inferior colliculus. *Journal of Neurophysiology*, **57**, 1130–1147.

Semple, M. N. and Kitzes, L. M. (1987b) Hierarchical processing of binaural level cues in the ascending auditory system. *Society for Neuroscience Abstracts*, **13**, 1467.

Sharp, F. R., Ryan, A. F., Goodwin, P. and Woolf, N. K. (1981) Increasing intensities of wide-band noise increase [^{14}C] 2-deoxyglucose uptake in gerbil central auditory structures. *Brain Research*, **230**, 87–96.

Shaw, E. A. G. (1982) 1979 Rayleigh medal lecture: The elusive connection. In R. W. Gatehouse (ed.) *Localization of Sound: Theory and Applications*. Amphora, Groton, pp. 13–29.

Smith, D. E. and Moskowitz, N. (1979) Ultrastructure of layer IV of the primary auditory cortex of the squirrel monkey. *Neuroscience*, **4**, 349–359.

Snowdon, C. T. (1988) Communications as social interaction: its importance in ontogeny and adult behavior. In D. Todt, P. Goedeking and D. Symmes (eds) *Primate Vocal Communication*. Springer, Berlin, pp. 108–122.

Sokoloff, L., Reivich, M., Kennedy, C., Des Rosiers, M. H., Patlak, C. S., Pettigrew, K. D., Kakaurda, O. and Shinohara, M. (1977) The deoxyglucose method for the measurement of local cerebral glucose utilization: Theory, procedure and normal values in the conscious and anesthetized albino rat. *Journal of Neurochemistry*, **28**, 13–36.

Sovijärvi, A. R. A. (1973) *Single unit responses to complex and moving sounds in the primary auditory cortex of the cat*. Dissertation, Institute of Physiology, University of Helsinki.

Sovijärvi, A. R. A. (1975) Detection of natural complex sounds by cells in the primary auditory cortex of the cat. *Acta Physiologica Scandinavica*, **93**, 318–335.

Sovijärvi, A. R. A. and Hyvärinen, J. (1974) Auditory cortical neurons in the cat sensitive to the direction of sound source movement. *Brain Research*, **73**, 455–471.

Spoendlin, H. (1972) Innervation densities of the cochlea. *Acta Otolaryngologica*, **73**, 235–248.

Starr, A., Amlie, R. N., Martin, W. H. and Sanders, S. (1977) Development of auditory function in newborn infants revealed by auditory brainstem potentials. *Pediatrics*, **60**, 831–840.

Steffen, H., Simonis, C., Thomas, H., Tilleim, J. and Scheich, H. (1988) Auditory cortex: multiple fields, their architectonics and connections in the mongolian gerbil. In J. Syka and R. B. Masterton (eds) *Auditory Pathway*. Plenum, New York, pp. 223–228.

Steinschneider, M., Arezzo, J. and Vaughan, H. G., Jr (1982) Speech evoked activity in the auditory radiations and cortex of the awake monkey. *Brain Research*, **252**, 353–365.

Strominger, N. L. (1969) Localization of sound in space after unilateral and bilateral ablation of auditory cortex. *Experimental Neurology*, **25**, 521–533.

Suga, N. (1965) Functional properties of auditory neurones in the cortex of echolocating bats. *Journal of Physiology (London)*, **181**, 671–700.

Suga, N. (1982) Functional organization of the auditory cortex. In C. N. Woolsey (ed.) *Cortical Sensory Organization*. Vol. 3. *Multiple Auditory Areas*. Humana, Clifton, New Jersey, pp. 157–218.

Suga, N. and Jen, P. H.-S. (1976) Disproportionate tonotopic representation for processing CF-FM signals in the mustache bat auditory cortex. *Science*, **194**, 542–544.

Suga, N. and Manabe, T. (1982) Neural basis of amplitude-spectrum representation in auditory cortex of the mustached bat. *Journal of Neurophysiology*, **47**, 225–255.

Suga, N., Niwa, H., Taniguchi, I. and Margoliash, D. (1987) The personalized auditory cortex of the mustached bat: Adaptation for echolocation. *Journal of Neurophysiology*, **58**, 643–654.

Sullivan, W. E. and Konishi, M. (1984) Segregation of stimulus phase and intensity coding in the cochlear nucleus of the barn owl. *Journal of Neuroscience*, **4**, 1787–1799.

Symmes, D. and Biben, M. (1988) Conversational vocal exchanges in squirrel monkeys. In D. Todt, P. Goedeking and D. Symmes (eds) *Primate Vocal Communication*. Springer, Berlin, pp. 123–132.

Theurich, M., Müller, C. M. and Scheich, H. (1984) 2-Deoxyglucose accumulation parallels extracellularly recorded spike activity in the avian auditory neostriatum. *Brain Research*, **322**, 157–161.

Thompson, G. C. and Cortez, A. M. (1983) The inability of squirrel monkeys to localize sound after unilateral ablation of auditory cortex. *Behavioral Brain Research*, **8**, 211–216.

Thompson, R. F. (1960) Function of auditory cortex of cat in frequency discrimination. *Journal of Neurophysiology*, **23**, 321–334.

Tomlinson, R. W. W. and Schwarz, D. W. F. (1988) Resolution of components of harmonic complex tones by single neurons in the alert auditory cortex. In J. Syka and R. B. Masterton (eds) *Auditory Pathway*. Plenum, New York, pp. 245–249.

Tunturi, A. R. (1952) A difference in the representation of auditory signals for the left and right ears in the iso-frequency contours of the right middle ectosylvian cortex of the dog. *American Journal of Physiology*, **168**, 712–727.

von Békésy, G. (1960) *Experiments in Hearing*. McGraw-Hill, New York.

von Economo, C. (1929) *The Cytoarchitectonics of the Human Cerebral Cortex*. Oxford University Press, Oxford.

Vrensen, G., De Groot, D. and Nunes-Cardozo, J. (1977) Postnatal development of neurons and synapses in the visual and motor cortex of rabbits: a quantitative light and electron microscopic study. *Brain Research Bulletin*, **2**, 405–416.

Walker, A. E. (1938) *The Primate Thalamus*. University of Chicago Press, Chicago.

Walsh, E. G. (1957) An investigation of sound localization in patients with neurological abnormalities. *Brain*, **80**, 222–250.

Warr, W. B. and Guinan, J. J., Jr (1979) Efferent innervation of the organ of Corti: Two separate systems. *Brain Research*, **173**, 152–155.

Watanabe, T. (1972) Fundamental study of the neural mechanism in cats subserving the feature extraction process of complex sounds. *Japanese Journal of Physiology*, **22**, 569–583.

Watanabe, T., Yanagisawa, K., Kanzaki, J. and Katsuki, Y. (1966) Cortical efferent flow influencing unit responses of medial geniculate body to sound stimulation. *Experimental Brain Research*, **2**, 302–317.

Webster, K. E. (1961) Cortico-striate interrelations in the albino rat. *Journal of Anatomy*, **95**, 532–544.

Webster, K. E. (1965) The cortico-striatal projection in the cat. *Journal of Anatomy*, **99**, 329–337.

Webster, W. R. and Aitkin, L. M. (1971) Evoked potential and single unit studies of neural mechanisms underlying the effects of repetitive stimulation in the auditory pathway. *Electroencephalography and Clinical Neurophysiology*, **31**, 581–592.

Webster, W. R., Servière, J., Batini, C. and Laplante, S. (1978) Autoradiographic demonstration with [^{14}C]-2-deoxyglucose of frequency selectivity in the auditory system of cats under conditions of functional activity. *Neuroscience Letters*, **10**, 43–48.

Weinberger, N. M. and Diamond, D. M. (1987) Physiological plasticity in auditory cortex: rapid induction by learning. *Progress in Neurobiology*, **29**, 1–55.

Wenthold, R. J. (1985) Glutamate and aspartate as neurotransmitters of the auditory nerve. In D. G. Drescher (ed.) *Auditory Biochemistry*, Chap. 8. Thomas, Springfield, pp. 125–140.

Wernicke, C. (1874) *Der aphasische Symptomencomplex*. Breslau, Cohn and Weigert, pp. 1–72.

Whitfield, I. C. and Evans, E. F. (1965) Responses of auditory cortical neurons to stimuli of changing frequency. *Journal of Neurophysiology*, **28**, 655–672.

Winer, J. A. (1984a) Anatomy of layer IV in cat primary auditory cortex (AI). *Journal of Comparative Neurology*, **224**, 535–567.

Winer, J. A. (1984b) The pyramidal neurons in layer III of cat primary auditory cortex (AI). *Journal of Comparative Neurology*, **229**, 476–496.

Winer, J. A. (1984c) The non-pyramidal cells in layer III of cat primary auditory cortex (AI). *Journal of Comparative Neurology*, **229**, 512–530.

Winer, J. A. (1985a) The medial geniculate body of the cat. *Advances in Anatomy, Embryology and Cell Biology*, **86**, 1–98.

Winer, J. A. (1985b) Structure of layer II in cat primary auditory cortex (AI). *Journal of Comparative Neurology*, **238**, 10–37.

Winer, J. A. (1986) Neurons accumulating (^{3}H) gamma-aminobutyric acid (GABA) in supragranular layers of cat primary auditory cortex (AI). *Neuroscience*, **19**, 771–793.

Winguth, S. D. and Winer, J. A. (1986) Corticocortical connections of cat primary auditory cortex (AI): Laminar organization and identification of supragranular neurons projecting to area AII. *Journal of Comparative Neurology*, **248**, 36–56.

Winter, P. and Funkenstein, H. H. (1973) The effect of species-specific vocalization on the discharge of auditory cortical cells in the awake squirrel monkey (*Saimiri sciureus*). *Experimental Brain Research*, **18**, 489–504.

Winter, P., Ploog, D. and Latta, J. (1966) Vocal repertoire of the squirrel monkey (*Saimiri sciureus*), its analysis and significance. *Experimental Brain Research*, **1**, 359–384.

Wise, L. Z. and Irvine, D. R. F. (1985) Topographic organization of interaural intensity difference sensitivity in deep layers of cat superior colliculus: Implications for auditory spatial representation. *Journal of Neurophysiology*, **54**, 185–211.

Wong, D. and Kelly, J. P. (1979) Laminar connections in the cat's auditory cortex (AI). *Society for Neuroscience*, **9**, 33.

Wong, D. and Shannon, S. L. (1988) Functional zones in the auditory cortex of the echolocating bat, *Myotis lucifugus*. *Brain Research*, **453**, 349–352.

Woods, J., Plessinger, M. and Mack, C. (1983) Fetal auditory brainstem evoked response. *Pediatric Research*, **18**, 83–85.

Woolsey, C. N. (1960) Organization of cortical auditory system: A review and a synthesis. In G. L. Rasmussen and W. F. Windle (eds) *Neural Mechanisms of the Auditory and Vestibular Systems*. Thomas, Springfield, pp. 165–180.

Woolsey, C. N. and Walzl, E. M. (1942) Topical projection of nerve fibers from local regions of the cochlea to the cerebral cortex of the cat. *Bulletin of the Johns Hopkins Hospital*, 71, 315–344.

Woolsey, C. N. and Walzel, E. M. (1944) Topical projection of the cochlea to the cerebral cortex of the monkey. *American Journal of Medical Science*, **207**, 685–686.

Woolsey, C. N. and Walzl, E. M. (1982) Cortical auditory area of *Macaca mulatta* and its relation to the second somatic sensory area (Sm II). In C. N. Woolsey (ed.) *Cortical Sensory Organization*. Vol. 3. *Multiple Auditory Areas*. Humana, Clifton, New Jersey, pp. 231–256.

Yin, T. C. T., Chan, J. C. K. and Irvine, D. R. F. (1986) Effects of interaural time delays of noise stimuli on low-frequency cells in the cat's inferior colliculus. I. Responses to wideband noise. *Journal of Neurophysiology*, **55**, 280–300.

Index